What's These Worlds Coming To?

Stefanos Geroulanos and Todd Meyers, *series editors*

What's These Worlds Coming To?

Jean-Luc Nancy and Aurélien Barrau

Translated by Travis Holloway and Flor Méchain

FORDHAM UNIVERSITY PRESS
NEW YORK 2015

This book was originally published in French as Aurélien Barrau and Jean-Luc Nancy, *Dans quels mondes vivons-nous?* © Éditions Galilée, 2011.

Cet ouvrage a bénéficié du soutien des Programmes d'aide à la publication de l'Institut Français.

This work, published as part of a program of aid for publication, received support from the Institut Français.

Library of Congress Control Number: 2014947278

Printed in the United States of America

17 16 15 5 4 3 2 1

First edition

CONTENTS

FOREWORD

To Inhabit a World

David Pettigrew

One is able to discern a certain trajectory in the work of Jean-Luc Nancy, from a thinking of *community* to a thinking of *world*, a trajectory that can be said to begin with his text *The Inoperative Community*,[1] which first appeared in French in 1986, and which culminates in this new book, *What's These Worlds Coming To?* [Dans quels mondes vivons-nous?].[2] Significantly, in the movement of his thought from community to world, Nancy has nonetheless remained concerned with the theme of *being-with*, which he has drawn from Heidegger's existential analysis of *Mitsein*.[3] In *The Inoperative Community*, Nancy cautions that our being-with others "is not a communion . . . nor even a communication as this is understood to exist between subjects. But these singular beings are themselves constituted by sharing, they are distributed and placed, or rather *spaced*, by the sharing that makes them *others*" (*IC* 25). These singular others, he writes, communicate by "*not* 'communing'" (ibid., my emphasis). The communication of "sharing" takes place in this "very dis-location" (ibid.). Any community in this sense would be composed of singular existences that "share" the exposure of their singularity in their being-toward-death, or finitude. What is "communicated" is nothing other than the exposition of singularity. For Nancy, "community" means that there is "no singular being without another singular being" (*IC* 28). Nancy writes, "This exposure, or this exposing-sharing, gives rise, from the outset, to a mutual interpellation of singularities prior to any address in language (though it gives this latter its first condition of possibility). Finitude compears, that is to say it is exposed: such is the essence of community" (*IC* 29).

What is shared, however, in this paradoxical sense, is the impossibility of sharing. Singular beings are given, he writes, "without a bond *and* without a communion" (*IC* 29). For Nancy, the being-with or "being-in-common" of an inoperative community is a community that, as such, can never cohere. What we might refer to as the *in-coherence of community* is crucial for Nancy's thinking. "At every instance," Nancy writes, "singular beings share their limits, share each other on their limits" (*IC* 41). Nancy opposes, then, the thinking of the limit of community, or the *in-coherence of community*, to an "absolute immanence" of community. For Nancy, it is just such an immanence that is "the stumbling block to a thinking of community," since it coheres in its immanence as a "totalitarianism," a totalitarianism that Nancy's thinking of an inoperative community seeks to avoid (*IC* 3).

In his text *Being Singular Plural* (2000), which originally appeared in French in 1996, Nancy advances a thought of a world that springs forth from a plurality of singular origins "everywhere and in each instant."[4] "The origin of the world," he asserts, "occurs at each moment of the world. It is the *each time* of Being, and its realm is the *being-with* of each time with every [other] time. The origin is for and by way of the singular plural of every possible origin" (*BSP* 83). For Nancy, the world is composed of the singularity of the plurality of primordial beginnings: "Each being belongs to the (authentic) origin, each is originary (the springing forth of the springing forth itself) . . ." (ibid.). The springing forth of the singularities that is at the origin of the world involves, as in the case of *The Inoperative Community*, a sharing. There is a sharing of origins that, for Nancy, is a thinking of *being-with*, a sharing that is intercorporeal. The "intercorporeal," Nancy writes, "exposes bodies according to their being-with-one another . . . *amongst themselves* [entre eux] as origins" (*BSP* 84). One mode of this exposition is language. Yet such an exposition does not entail a prosaic mode of communication. What language expresses "is the exposing of plural singularity" (ibid.). In language, "all of being is exposed as its meaning [*sens*] . . . as the originary sharing according to which a being relates to a being, the circulation of a meaning of the world [*sens du monde*] that has no beginning or end" (*BSP* 84). The relation of singular beings is, "each time, the punctuality of a 'with' that establishes a certain origin of meaning [*sens*]" (*BSP* 85).

This coexposition of the *body* and *language* is further articulated in Nancy's treatment of the *body* and *writing* in *Corpus*.[5] For Nancy the "*exscription* [excription]" of the body entails a "being placed *outside the text* as the most *proper* movement of its text (*C* 11). He writes, "we have to write from a body that we neither have nor are, but where being is exscribed" (*C* 19). In the process of writing we are undone and we "lose of our footing," since, in the exscription we are caught up in the intercorporeality of expression (*C* 13). Nancy writes that there is "*no writing that doesn't touch*" (*C* 11), and in this touching there is a "*breakthrough* [effraction]" as the body "*exposes a breakthrough of sense*" (*C* 25). Nancy addresses this conceptual breakthrough proper to the body with the neologism "*expeausition*," a term that replaces the phoneme "*po*" (in ex*po*sition) with the homonymically equivalent French word for skin, "*peau*" (ex*peau*-sition) (*C* 33).

For Nancy, then, the thinking of the origin and sense of the world, in *Being Singular Plural*, entails a thinking of *being-with*. Indeed, the *inter*corporeal exposition of language "exposes the world and its proper being-with-all-beings in the world" (*BSP* 85). It is nothing less than this intercorporeal ex-position that makes "the world 'hold' or 'consist' in its proper singular plurality" (ibid.).[6] Thus, Nancy's thinking of community in *The Inoperative Community*, in the sense that it involves the sharing-out of singular beings in their limits or finitude, is intertwined with his thinking of world in *Being Singular Plural*. Moreover, *Being Singular Plural*, with its thinking of the exposition of world on the basis of a being-with, can be said to serve as a *hinge-work* between Nancy's thinking of community and his thinking of world.

With his text *The Sense of the World* (1997), which originally appeared in French in 1993, Nancy identifies a "becoming worldwide [*mondialisation*] of the world," a "cosmopolitanism" and "teletechnism" that is tearing the sense of the world "to shreds."[7] Nancy seeks to salvage the sense of the world, grasping at the "only chance for sense and its only possible sense" beyond the structures and restrictions of grammar in the "abandonment of sense, as the opening of the world" (*SW* 3). Nancy is engaged with this loss of the sense of the world as he writes: "[T]he world of sense is culminating today in the unclean [*l'immonde*][8] and in nonsense. It is heavy with suffering, disarray, and revolt" (*SW* 9). For Nancy, to make sense of the sense of the world, to "sense oneself making sense [*se sentir faire sens*], and even more, to sense *oneself as the engenderment of sense* [se *sentir* comme l'engendrement du sens] . . . is without a doubt the ultimate stake of philosophy" (*SW* 162). To make sense of the world means to encounter the origin "where it [*ça*] opens itself" (*SW* 160). In other words, Nancy insists on the singularity of the opening or origin of the world that is always already in excess of any other or any previous meaning. This excess is expressed in "the exscription of all words: the taking-place-there of their sense, of all their senses, 'outside,' here" (ibid.). Making sense would realize this intercorporeal dehiscence of sense "through the common being-as-act of sensing and sensed" (*SW* 78). The sense of the world would emerge as "a differentiated articulation of singularities that make sense in articulating themselves along the edges of their articulation" (*SW* 78). Nancy points, indeed, to an "active dehiscence of the act of sensing: that is to say, to *ek-sisting* in general" (*SW* 79).

In his text, *The Creation of the World* or *Globalization* (2007), which first appeared in French in 2002, Nancy addresses the sense of the world in terms of the crisis of globalization. He provides a sharp distinction between "globalization," on the one hand, and an authentic "world-forming," on the other hand. Rejecting the *un-world* [l'immonde], Nancy emphasizes the freedom of singular plural beginnings when he writes, "*To create the world* means: immediately, without delay, reopening each possible struggle for a world, that is, for what must form the contrary of a global injustice against the background of general equivalence" (*CW* 54). The suppression of the creation of meaning, of "each possible struggle for a world," would, for Nancy, constitute injustice. To the "un-world" of technology wielded by metaphysics, Nancy opposes a world that is always under formation. For Nancy, "What *forms a world* today is exactly the conjunction of an unlimited process of an eco-technological enframing *and* of a vanishing of the possibilities of forms of life and of common ground" (*CW* 95). Justice, for Nancy, would be engendered by the inexhaustible creation of meaning.

In our present text, *What's These Worlds Coming To?* Nancy addresses, among other things, the sense of the world in terms of the crisis of its technological enframing. His reflections may be read as a reference to what Heidegger named *das Ge-Stell* in the Bremen lectures.[9] For Heidegger, *das Ge-Stell* "essences as the plundering drive that orders the constant orderability of the complete standing reserve" (*BL* 31). Heidegger asserts that the standing reserve "persists in requisitioning [*Bestellen*]," as a "machination of the human, executed in the manner of an exploitation" (*BL* 28). Yet this machination is not a strategy carried out by humans but is rather one that envelops humans. Through the requisitioning of *das Ge-Stell*, humans and nature are "machined," in a sense, into pieces of the standing reserve, and rendered equivalent, interchangeable, and replaceable: "one piece can be exchanged for the other" (*BL* 35). In the ordered equivalence of the standing reserve, "everything stands in equal value" (*BL* 42). This interchangeability and replaceability of the standing reserve [*Bestand*] could be referred to, for example, as an alienation of *Dasein*, in this context, from its proper sense making.

For Nancy as well, the technological frames the world as in a machine. Moreover, the machine is not separate from the world but the world itself becomes a technologized machine. "Technology," he writes, "is a structuration of ends—it is a thought, a culture, or a civilization, however one wants to word it—of the indefinite construction of complexes of ends that are always more ramified, intertwined, and combined, but, above all, of ends that are characterized by the constant redevelopment of their own constructions" (*WTW* 44). In other words, the technologized machine, or apparatus, has no end but its own end, an end without sense and without social value, an end that leads to market volatility or to the destruction of worlds as in the case of genocide.[10] Within the machine is a hypertrophic construction that makes it less and less possible to distinguish between subject and object, human, nature and world, entailing moreover a loss of agency, responsibility, and sense. All these elements are dispersed out into the machine, into what Nancy identifies as an "*ecotechnology* that our ecologies and economies have already become" (*WTW* 54). What is at stake in this ecotechnology, for Nancy, is nothing less than the sense of the world. The questioning of the sense of the world offered in this new book, then, takes its place in Nancy's ongoing inquiry, as he writes, "The sense of 'world' is not only undecided and multiple—it has become the crucial point where all of the aspects and stakes of 'sense' in general become tied together" (*WTW* 1).

Perhaps in the context of *What's These Worlds Coming To?* the "being-with" is enacted in the *co*-authorship of the text with Aurélien Barrau, who writes alternating chapters. Further, the *with* of a being-with, or a *living-*with, appears in a recent interview with Pierre Philippe Jandin, titled *La possibilité d'un monde: dialogue avec Pierre-Philippe Jandin.*[11] In the interview, Jandin asks Nancy about his collaboration with Barrau. Nancy replies that while Barrau is an astrophysicist, he is "also a philosopher" (*PM* 34). Nancy states, moreover, that he has learned from Barrau that astrophysics is "obliged to give thought to a plurality of worlds," that is, to a "*plurivers*" or a "*multivers*" as opposed to a "*univers*" (ibid.). This reference to a plurality of worlds offers some perspective on the meaning of the new book's enigmatic title (*Dans quels mondes vivons-nous?*). Perhaps *being-with* can still be heard in the French title of the text, that is, through the "*we*" [nous], if the title is translated literally as "What worlds do *we* live in?" Nancy suggests in *La possibilité d'un monde* . . . that we need to learn to inhabit the world anew, "to be in the world" [*in-der-Welt-sein*] necessarily engaged in the circulation of sense that makes a world (*PM* 31). Such a "*habitus*," he states, "is not far from an *ethos*," and he insists, moreover, that "what we need is an ethics of the world" (ibid.). Nancy asserts that "The world is the possibility of circulation of sense and *we* have to make a world, and remake a world" (ibid., my emphasis).

Nancy's gesturing toward an "ethics of the world" is evocative of his essay titled "Heidegger's 'Originary Ethics.'"[12] In that context, Nancy referred to a "thinking" that is the "experience of this absolute responsibility for *sense*" (*HPP* 80, my emphasis). For example, absolute responsibility would entail "opening oneself to making sense as such" (*HPP* 81). For Nancy, an originary *ethos* would entail a dwelling with a responsibility for making sense and for a possibility for making sense (cf. *HPP* 82). This responsibility for sense making would be an origin of a world to be, as Nancy suggested, made and remade. In the context of an originary ethics, Nancy refers to a "making sense in common," which is nothing less than "finitude as sharing" (*HPP* 83). A finitude in sharing is what Nancy has referred to as a being singular plural of a making sense that is at the origin of the world. Indeed, in a 2008 essay titled "The Being-with of the Being-There," Nancy asserts that "*Dasein* is a singular, unique possibility of forming/letting a proper meaning of the world . . . open."[13] Perhaps what Nancy undertakes in that text is not so much a strict interpretation of Heidegger's treatment of *Dasein* and being-with, as it is a palpation of the possibilities and the stakes of being-with:

"[I]t is exactly at the site of the *with* that both the opportunity and risk of existence are manifest" (*RF* 126).

In the "Preamble" of this new book, Nancy and Aurélien raise questions with respect to the sense of the *world*. Specifically, the authors identify a certain threefold crisis. They state, first, that in our time the cosmos can no longer be represented as a coherent whole or unity. Second, they state that the world lacks a definite and manageable order according to which we would address nature or culture. Third, they suggest that the world has become diversified and multiple as never before, affecting the complexity of our interactions (with life, matter, space, and time), and destabilizing all forms of civilization (cf. *WTW* 1). The authors lament that in a so-called globalized world, our world and our modes of life are "more diffracted, scattered, heterogeneous, and even unidentifiable" (*WTW* 3). It is in this context that Nancy gives thought to the sense of the world in crisis. In Chapter 1, his approach to this question is to give thought to the "*plus d'un*" of the world, an expression borrowed from Derrida that, in this context, means or suggests "more than one." Nancy clarifies that this "more than one" is not "more" in the sense of adding one to another in a countable series. Rather, for Nancy, the "more than one" hearkens to Derrida's thinking of "dissemination" and "*différance*," in order to encounter the meaning or sense of the world as an "excess of sense" (*WTW* 12). Nancy speaks of such a *différance*, or such an excess, as "*l'un excédant*," or "the exceeding one" (*WTW* 18). The world, then, "is," for Nancy, an excess of sense. Nancy's thinking of a world as exceeding or preceding itself is an opening of a critique, on the one hand, of the inappropriateness of the scientific view of the world as a unified object in space with measurable dimensions, or as a container that we are "in," and, on the other hand, a critique of the commercial tendency to objectify and render everything as "merchandise" within a globalized world. Nancy's thinking of a world as exceeding itself is an opening to "sense" itself.

Hence, in the course of his work, Jean-Luc Nancy has undertaken an interrogation of the sense of the world that began with a thinking of community and continued with a thinking of the world. In the course of this trajectory, Nancy has developed a thought of being-with and being-in-common. In the current book, one encounters a sense of being-with in terms of a living-with in a world. But for those readers who still have "a taste for the secret,"[14] I will not say more about the current book, but instead offer the following passage from the text, to give Jean-Luc Nancy and Aurélien Barrau the last word:

> We do not conclude: We open up, we attempt. *Dans quel(s) monde(s) vivons-nous?*—the ambiguity of whether it is a pluralized singular or a singularized plural is admitted—may be heard in the same key as: "Is there something like a world in which we find ourselves?" Or perhaps: "Of all the possible worlds, which one is ours?" Or might we even understand the "world" in the sense of when one cries out: "The whole world is here!" By which crowd, which pack, which fray are we being carried away? Can one even identify, distinguish, and disentangle such a web?
>
> Whatever the case may be, today one must get in the fray. (*WTW* 7)

TRANSLATORS' PREFACE

Droughts and floods. Stagnant wages and unpayable mortgages. Modified food. "Socialist" presidents! "Citizenship" and "marriage" for everyone?! The world as we know it is crumbling. What will replace it seems more uncertain than ever. How did it come to this?! *Dans quels mondes vivons-nous?! What's these worlds coming to?*

The original, untranslatable title of this book in French—*Dans quels mondes vivons-nous?*—is odd, to say the least. The title appropriates, distorts, and deliberately misuses a common expression that is used to comment on the state of the world. The phrase *dans quel monde vivons-nous* or *what's this world coming to* expresses a range of emotions about the world from despair and resignation to outrage and revolt. Yet the coauthors of this book—a leading French philosopher and an award-winning philosopher-astrophysicist—do not simply repeat this common idiom in the title. They alter it by adding a silent *s* so that the phrase *what's this world coming to* or *dans quel monde vivons-nous* becomes something akin to *what's these worlds coming to* or *dans quels mondes vivons-nous*. The pronunciation of both expressions is the same in French.

While the addition of this silent *s* is very strange, it is not unprecedented. One cannot help but remember how one of Nancy and Barrau's major interlocutors, Jacques Derrida, chose to misappropriate and misspell *différance* with a silent *a* (and stories of Derrida's mother rightly scolding him upon hearing about this: "Jackie, you didn't spell difference with an *a*, did you?"). Beyond this, one might think of an entire history in which, during moments of great change or rupture, thinkers and artists and scientists found it necessary to propose a shift in common sense. In such a time as this, Barrau and Nancy's misappropriation of the idiom *what's this world coming to* is a deliberate interruption and inscription—an interscription—that sets out to rethink our lives from the everyday supermarket to the most distant stars.

The challenge, Barrau and Nancy tell us, is that none of our old language or idioms refer to a new world that has quietly cropped up on us and is, in fact, already here. We no longer live in a world, but in worlds. We do not live in a larger universe, but in a multiverse. We no longer create; we appropriate and montage. And finally, we do not build sovereign, hierarchical political institutions anymore; we form local assemblies and networks of cross-national assemblages at the same time as we form multinational corporations that no longer pay taxes to the state. We find ourselves *living* among heaps of odd bits and pieces that amass without any unifying force or center, living in a time not only of ruin and fragmentation but also of rebuilding. In this time of rebuilding, philosophical thought has shifted from deconstruction to *struction*. And while this does not mean a return to first philosophy's *archē* and order, it also does not mean that thought has stopped at the deconstruction of it. Contemporary thought is now focused rather on what Barrau and Nancy carefully call the *struction* of *dis-order*.

For all of these reasons, *what's these worlds coming to* is not only an idiom expressing resignation or outrage about the state of the world. It is a line of inquiry and a site of *struction*. After all, what kind of worlds *are* we living in? What is *life* or *living* in *these worlds*? What's these worlds coming to, indeed?

These multiple new worlds are already here and still undecided. But our beliefs and languages are not even able to address them. It is curious, then, that Barrau and Nancy's deliberate misappropriation of idiomatic language is not limited to the title in this work. The gesture is repeated throughout a text that attempts not only to deconstruct our clichés but also to begin to resew or restitch the way we talk about the cosmos. Nancy, for instance, breaks the rules of grammar when he finds his language incapable of addressing the kind of plurality that he sees coming. Barrau misappropriates a well-known poem by Éluard ("Liberty") only to evade this poem in the end. Both weave together seemingly idiomatic phrases that are built from pieces of different, existing idioms. The phrase *plier voile*, for example, collages and juxtaposes two typical French expressions, *plier baggage* and *mettre les voiles*, both of which mean "to leave discreetly." Such gestures go beyond puns and venture into *détournement*, malapropism, and reappropriation. Still, phrases such as "take back the terms" or "reappropriate the terms" show us that Nancy and Barrau do not seek a pure and total departure from the world but rather a participation with it and a rupture within it.

This sort of play with language will feel strange to many English readers at times—as if one were in a bizarre Max Ernst collage or a Giorgio de Chirico cityscape that exists everywhere and nowhere. One may recall, however, that the tradition of philosophers and astrophysicists has indeed always been a queer one. From the questioning of our habits in Aristotle's *Nicomachean Ethics* to the discussion of *order-words* by Michel Foucault, philosophy itself could be defined as the struggle to propose a sense that is within and alongside common sense but one that is also radically different from it. Likewise, when astronomers such as Galileo or Bruno showed, for instance, that the earth was not the center of the universe or that multiple worlds existed, these wonderers were burned at the stake or forced to renounce their work by an all too *catholic* church whose answer to the question of life in this world was absolute: We live in a harmonious universe, created and set into motion by a first mover, directed toward a final aim or end, endowed with sovereign representatives or monarchs whose power and laws and institutions are manifestations of the divine order, and whose primary edict is eternally clear: maintain order—maintain the order of the uni-verse.

Thankfully, Nancy and Barrau interrogate and celebrate a very different world or worlds beginning with their curious title. A few points should be noted, however, about the English translation of this idiom. The meaning or meanings of an idiom, by virtue of its being an idiom, is of course separate from the literal or stable meanings of the words that occur in the idiom—a point Derrida discusses at length in *The Truth in Painting*. Thus, a literal translation of the distorted idiom in the original title—"what worlds are we living in"—fails to capture the panoply of meanings in the original French idiom, the palpable affects that arise when this string of words is used, and even the intentional use of everyday or idiomatic language by academics.

Unfortunately, two very important words, however, are lost in translation in the distorted English idiom *what's these worlds coming to*. First, the pronoun "we" in the original French title will be familiar to readers of Nancy. Nancy and Barrau extend Nancy's notion of the "we" or "singular plural" throughout this work—not in anthropomorphic or exclusively human terms, but rather in terms of new ensembles of the living or of "all of the living together." This second untranslated word, "living," also implies a longer conversation that precedes it—one that is both very ancient and very contemporary. Much like Plato's *Timaeus*, in which the discussion of the cosmos or world is linked to a discussion of the body, Nancy and Barrau refer to these worlds as pulses, phalanxes, dactyls, vertebrae, nerves, or the "fertile push of the living." A more recent discussion of "life" or "living" can be found in contemporary philosophy, for example, in the work of Derrida (*Learning to Live Finally*), Gilles Deleuze ("Immanence: A Life"), or Giorgio Agamben (*Homo Sacer: Sovereign Power and Bare Life*). Much more could be said about this.

Second, it is important to note that the phrase "coming to" does not appear in the original French idiom, although this language appears frequently throughout the text. This particular "coming to," however, should not be understood as a direction toward an Aristotelian goal or *telos*. And yet it must also not be understood as the *à venir* or "to come" of Derrida—that is, as worlds that are always deferred and never present and therefore always still *to come*. Nancy and Barrau's "coming to" would be closer, in fact, to Nietzsche's word *heraufkommen* or *coming up*, a coming or arising that is coming up from or out of [*herauf-*] the conditions of the present world. This particular "coming to," then, is already here and still arriving. Already under way or in process and yet undecided because it is still taking place or rising up in the life of these worlds. It is what Nancy and Barrau call the *survenir*, the "coming about" that is already "in the works . . . beneath our eyes and in our words." As Nancy writes, this *survenir* of struction "opens less onto a past or future and more onto a present that is never really accomplished in presence."

Finally, the translators regret to note that *what's these worlds coming to* fails to capture the silent *s* in the original idiom—a letter that is written and read in the original title, but not spoken or heard.

Our difficulty in translating all of these meanings might be compared to an old story about another philosopher and astronomer that goes something like this. A long time ago, a curious astronomer took a walk at night to look up at the stars and fell down into a deep well. Although someone apparently mocked the astronomer for this, it is uncertain whether the story is a comic joke about the hubris of the astronomer or a tragic foreboding about his fate, a false, stubborn cliché about learning, or a frank, outspoken truth about education. It might even be a myth about the underworld, in which case, just as Socrates went down to Piraeus and Empedocles leaped down into fire, so Thales fell down into water. It is not clear who the author of the story is and, depending on which version of the story is told, it is not clear whether the man in question is an astronomer or a philosopher or both. Rather than outlining the so-called origin of philosophy—the story of Thales, the earliest known Greek philosopher, who is said to have thought that the entire world is based on the unary *archē* (origin, principle, or sovereignty) of water—the story shows itself to have many undecided meanings or interpretations. Nancy and Barrau risk ambiguities such as this in order to invite us on an uncharted walk into barely known worlds. Not a tour around the world, but an astronomical survey that ends up at a pass where they make a choice to the leap into the well. *For heavens' sake.*

We wish to acknowledge the help and support of John Ashbery, Ron Avitzur, Aurélien Barrau, Benjamin Downs, Marc Dubois, David Espinet, Spencer Everett, Stefanos Geroulanos, Thomas Lay, Geoffrey Lind, Therese Malhame, Todd Meyers, Marie-Eve Morin, Michael Naas, Jean-Luc Nancy, Eric Newman, Anne O'Byrne, David Pettigrew, and François Raffoul. This translation would not have been possible without Helen Tartar.

Travis Holloway and Flor Méchain
New York and Paris
Summer 2013

Even as science progresses, however, it convinces us that we are becoming less and less capable of mastering by means of thought phenomena that, by their spatial and temporal orders of magnitude, escape our mental capacities. In that sense, the history of the cosmos is becoming a kind of great myth for the ordinary mortal: It consists of the unfolding of unique events whose reality, because the events occurred only once, can never be proven.

It has been possible since the seventeenth century to believe that scientific thought stands in radical opposition to mythic thought and that one would soon eliminate the other. We may now wonder, however, whether we are not observing the beginning of a movement in the other direction. Does not the very progress of scientific thought push it toward history? That was already the case in nineteenth-century biology with the theory of evolution, and modern cosmology is also oriented in that direction. I have attempted to show that, even for us, historical knowledge preserves affinities with myths. And if, as it seems, science itself is tending toward a history of life and of the world, we cannot rule out the possibility that, after long following diverging paths, scientific thought and mythic thought will one day move closer together.

—CLAUDE LÉVI-STRAUSS, *Anthropology Confronts the Problems of the Modern World*[1]

Preamble

I

The world—this term so broad and imprecise, while also polysemous—is undergoing three transformations of far-reaching importance. It can no longer be represented today as a "cosmos" (the ordering of a well-composed set or ensemble). It is now devoid of any manageable and definite order (both on a "universal" scale and on every level of "nature" and "culture"). And finally, it has been diversified and pluralized like never before, both in the complexity of our interactions with the given (matter, life, space, and time) and in the upheavals that affect all forms of civilization (knowledge, power, and values). For this threefold reason, the sense of "world" is not only undecided and multiple: It has become the crucial point where all of the aspects and stakes of "sense" in general become tied together.

At the same time, the world is confronted with three possibilities: three risks and three opportunities at the same time. It is entering into a movement of indefinite expansion, both on a "cosmic" scale and in our methods of knowing and acting on it or within it. It presents us with a diversity of issues that are themselves indefinitely multiplied as well (for example: Where does life begin? Can we still speak of a "nature"? A "matter"? A "history"? Or is it necessary to decompose, destabilize, and set apart these concepts?). Finally, we can no longer be certain of a distinction between "the world" and "us," between some thing that is in front of or around us and ourselves as "subjects" of this object. Perhaps there is no longer any room to speak about our being "in" the world like contents in a container. Perhaps instead we must learn the unique yet nonunified, universal yet multiversal existence of all ensembles and everything together [*tout ensemble*].

II

"What's this world coming to?": usually the question mark in this expression amounts to an exclamation point. The expression signifies: "What an unbearable and futile world we find ourselves in!" It strikes a chord of revolt and one of resignation at the same time. It belongs to the same school of expressions as "the world is falling apart" and "how did we get ourselves into this?" Within this use of the word—or idea—of "world" lies the strongest value that one could attach to it: that of the *cosmos*, the harmonious ensemble of celestial bodies whose orbs bear the relations of the universal order—that is, an order attuned to an integral unity. The sense and oscillation of this *order* and this *one* is what this question implicitly interrogates. To call into question the totalizing hegemony of this couple, which is also a movement, is the challenge with which we are confronted.

It so happens that the *cosmos* that is studied or invented by cosmophysics today no longer corresponds to this depiction of an absolute and unitary harmony. What metaphysics contemplates under the name world—in singular or in plural, as a "real" or "possible" world—probably corresponds to it even less. And the projection that we formerly believed could be made of the *cosmos* into an ordered history toward a particular end now seems radically discredited.

This is why physics and metaphysics are connected to one another from now on in a new way: the world, the experience of the world—the scientific as well as the existential experience of it—confounds the "cosmic" postulate to which we have become accustomed. "What's this world coming to?" is not merely a complaint or outcry anymore. It can also become a true line of inquiry. From now on, what can we, what must we, assign to this word that designates space–time, element, that which encompasses all of our presences, the "all together"?

On the one hand, the world–cosmos is fractured; on the other hand, the very idea of "world" (one, ensemble) no longer answers to the investigation of physics or to metaphysical questioning. "Pluriverse" and "multiverse" are now on physicists' agendas, while "multiplicity" and "multitude" permeate sociologies as well as ontologies. And yet, passing from unity to plurality cannot consist merely in multiplying exponentially, in increasing discrete numbers of unities (such as, for example, when we dwell on "multiculturality"). In such a transition or transformation, these are the paradigms both of the "one" and of the composition or construction of unities that are brought into play.

At a time when we say that every day the world is more "globalized" [*globalisé*] or "world-formed" [*mondialisé*][1]—and thus unified—and simultaneously that our worlds and our ways of life and culture are constantly more diffracted, scattered, heterogeneous, and even unidentifiable, the question of the world must be reconsidered: the question of the idea of "world," of the "reality" of the where in which we live, of its "unity" and "unicity," texture, or dissociation. We continue to hold that we live in one world even though it is not certain that we can still speak in these terms (either of "one world" or of "living"—although this would be a different consideration to take up altogether). In what terms would it be suitable, then, if not to speak of our existence (of its "sense" . . .) at least to get closer to the conditions of possibility for a speech that is suited to the "world" that is coming?

III

Such is the question that occasioned the meeting of the two signatories of this book.

Our meeting is not that of a physicist and a philosopher. Indeed, it is not even the creation of a graft or a hybrid. Rather it proceeds from a conjunction analogous to what other species of cosmologists or cosmosophers undertook in the past so as to transform necessity into fate. Still, this specific meeting does not concern necessity or contingency, nor does it create unity or disparity. We have disaffiliated, dispersed, and diffracted ourselves. Not "outside the field" or discipline, but precisely within, or under, what constitutes a true field.

The conjunction of our interests itself is the result of a shift that is taking place in the significations and stakes of what are called "science" and "philosophy." On either side, the assurance of an ordered truth of the universe (and within the universe) is becoming unclear. This is not a simple parallelism that would in turn refer back to a deeper order: It is not "the world" in itself that is evolving in such a way that it modifies our branches of knowledge and thought. Instead, it is that "the world" withdraws and transports in a spiraling manner the consistency of its reality "in itself." The world is neither real in the sense of an objective exteriority, nor unreal in the sense of a dream. More than "real" or "unreal," it is the carrying out of the interaction and the intertwining between "us" and "it."

Still this interaction is also the interaction from where we originate—both "us" and "the world." After all, we come from "nature," which in turn gets its consistency from us. This consistency is at times "physical," if we understand the word as Aristotle understands *phusis* (that which is brought about by itself), and at other times "natural," if we understand by natural that which is found outside of "culture." These significations—and those that might follow after it like variations on the world's "origins" or "ends" or "material" constitution—depend on our "culture"—a culture that is being transformed at the same time as the intimacy of our entry "into" this "matter" of the world is being transformed as well.

As we enter into it, we breach through [*faire irruption*] to the outside more and more: there precisely where there is no opposition between "inside" and "outside," between "world" and "nonworld." We are no longer in accord with a universal movement, nor are we in consideration of an object. We are neither "in" nor "in front of" the world anymore. We are the world and the world relates to itself in us and through us. This in no way means that we are the subjectivity of the world, but rather that the world is itself the "subject" of which we are an effect.

Still, this "subject" cannot be represented as a "relation to itself" without opening within itself the gap, the distance, and the inconsistency that is implied by any relation to oneself and to the other (if it is even possible to establish this difference). Thus, we are ourselves the inconsistency of the world with itself.

This gap presents itself in the language of our culture as the gap between physics and metaphysics. Yet it does not allow itself to be understood as an opposition between "science" and "speculation" or between "objectivity" and "subjectivity." On the contrary these distinctions crumble and each term passes from one into the other: matter cannot simply be allocated to the impenetrable, and spirit cannot simply be sublimated into the impalpable. This twofold movement, this dual erosion, possesses a point of intersection.

It is at this point, one could say, that we have met. At this point that is situated either nowhere at all or in countless places, in countless worlds.

IV

A slip [*glissement*] is taking place. It is in the works. It is at work—enacting, active—beneath our eyes and in our words. Some thing or some idea is withdrawing itself from us. An unhinging is taking place. As always, we are necessarily both its authors and its witnesses. We write it and read it in one and the same gesture. It is too soon to know or to sense whether it is a rupture or an evolution, whether the mode is that of a displacement or a break, whether the movement is reversible or absolutely oriented. The one-and-only-ordered-world is crumbling. Under different forms, through different ways, and for different reasons, a "pluriverse" is being invented–imposed and it remains essentially to be built *and* surveyed. Or is it perhaps rather, as astrophysicists like to call it, a multiverse, megaverse, or metaverse?

We follow this movement without letting ourselves be totally carried away by it. We have chosen in this book to approach it in the manner of a pendulum: Each swing of the pendulum would be autonomous in its effects and heteronomous in its manner of continuing, that is, of calling out for the next swing. Two positions that send off [*envoi*] and two proposals that send back [*renvoi*].

What first had to be understood was what this plural signifies. At least to get close to some of its essences: numeric or divisible, arithmetic or rhythmic. There are, evidently, several ways of not being one, of not being the One. Is it not precisely on a triptych that the primitive or primordial split in two must, *in fine*, be inscribed? Does the binary not omit the step [*pas*] itself: the initial as well as the initiated impetus? The step—the leap out of nothing or the leap of the nothing into something—is neither "one" nor "other" (anymore than it is "one" or "two"). It engages every "one" in something that is more and/or less than unity. If, therefore, we can no longer simply speak of the world in the singular, neither can we simply continue to affirm that there are several worlds. Rather the unity and unicity of the world follow a logic that is not that of the stability of the count (one, two, three, four . . .) but that of the dynamism or rhythm of what does not lend itself to being counted. This dynamism or rhythm is at once singular and plural, at once gathered and dispersed, at once contracting and expanding. Hence, perhaps we do not live in a world or in several worlds. Perhaps it is rather that the world or worlds unfold, diverge, or intersect in us and through us. Are we therefore distinct from the world? This is our first question.

This new plurality—this other way of naming and thinking a multiple that has always been subjacent, subtending—resonates, it seems to us, as a possible meeting place for analytic and continental philosophies. Is there fodder here for discovering if not a point of agreement at least a convergence between intellectual currents or circles that are both so distant in their methods—and sometimes in their wagers—and yet so close in the material that they create or dismantle to the point of contact, to the point of touching? The second movement of this work will also be the time to inquire about the proclaimed unity (actual or virtual) of the physical–mathematical description–construction of the world, and also to question its hegemony in light of—or in the shadows of—a primitive beat or pulsation. Here the myth of a real that is a unitary–unity finds itself undermined.

However, if this foundation, or the erosion that takes its place, must be comprehended in a plural mode, then in the third section of this work all of the constructivist logic, the articulation of *phusis* and *technē*, must be rethought beyond construction and deconstruction. Can we peer into [*entre-voir*] or (de) scribe [*(d)écrire*] the "all together" that is without order or disorder?

The all too reliable and well-oiled dialectic between the one and order is finally questioned in the fourth and final chapter. It is neither easy nor insignificant to jam this machine that has so carefully been assembled by a long philosophical and scientific tradition. The virgin earth—a strange territory or familiar non-place—of the dis-ordered multiple remains to be explored. Or invented.

V

We have tried to enter into this exploration or invention in a manner that is itself exploratory and tentative. Instead of constructing an ensemble, we arranged texts two by two: first, on the subject of unity itself; second, on the subject of construction—in other words, once on the subject of the differentiation of the "one" and once on the subject of its constitution. One of us (JLN) has put forward a text to which a text of the other (AB) responds.

We do not conclude: We open up, we attempt. *Dans quel(s) monde(s) vivons-nous?* [*What's this (these) world(s) coming to?*][2]—the ambiguity of whether it is a pluralized singular or a singularized plural is admitted—may be heard in the same key as: "Is there something like a world in which we find ourselves?" Or perhaps: "Of all the possible worlds, which one is ours?" Or might we even understand the "world" in the sense of when one cries out: "The whole world is here!" By which crowd, which pack, which fray are we being carried away? Can one even identify, distinguish, and disentangle such a web?

Whatever the case may be, today one must get in the fray.

More Than One

> We shall at one time be descending, tearing apart, like Osiris, the one into many by a titanic force; and we shall at another time be ascending and gathering into one the many, like the members of Osiris, by an Apollonian force.
>
> —PICO DELLA MIRANDOLA[1]

I

> [T]he double did not only add itself to the simple. It divided it and supplemented it. There was immediately a double origin plus its repetition. Three is the first figure of repetition. The last too, for the abyss of representation always remains dominated by its rhythm, infinitely. The infinite is doubtless neither one, nor empty, nor innumerable. It is of a ternary essence."
>
> —JACQUES DERRIDA, *Writing and Difference*[2]

One, two, and the resumption of this division and this addition. This resumption counts for one more, which makes three. The two divides the one and supplants and supplements it: the one has not taken place; it has only taken place by redoubling and repeating itself.

The "abyss of representation" is the nonpresence that is performed again with each new offer of signification: a sign, as soon as it signifies, refers back to another sign, and their connection connects or refers to nothing (to nothing as a "thing," a "presence," a "given"). Sign and sign and nothing—such is the rhythm.

And yet we are beginning to see that nothing [*rien*]—no thing, not any *one* thing [*auc*une], not a "one"—is "something" as it is implied in the French language (*res*, *rem*, a nothing [*rien*]).[3] In order to understand which "thing" is nothing, if that can even be understood, perhaps one must consider how it is referred back to or referenced: through rhythm. In other words, nihilism consists in holding that no sign refers back to a thing, and that signs only link together this nullity. It must be held, accordingly, that "a" corresponds to "one," to a "one." Still, if the repetition of the signs—language, and even more so, the significance of all things—functions as rhythm, is there not a scene change? Rhythm brings into play more than one. More than one, more than only one, "neither one, nor nil, nor innumerable" and *nothing*, the thing, in the form of the infinite. Exit nihilism. This is precisely what we are concerned with here.

The di- of division or the re- of repetition makes "one" more—one that is not another one added to the indefinite series of units, but one in which case its addition is also the gap between one and two. This gap divides one at the same time as it supplants and supplements it. In other words, it takes the place and the role of the one and in doing so it conjoins the one to itself—the simple "one" to the "one" that is split in two. This adjoining by division, this bending, this joint or articulation transforms the monotonous addition of one and one into rhythm. That is, linear succession is replaced or substituted by the return of the same—with this return Nietzsche opened what could be called the epoch of the "more than one." The "same," in effect, is not the "one"—at least, it is not identical to it. It is that for which being same or similar (*das Gleiche*) must specifically not be "one." Or rather, it must bring about the one—with its own unity and unicity—into a displacement, the movement of which we are beginning to feel, though in reality we sense that it is since the start of our history that the "one" displaces itself in itself or outside of itself. We will have reasons and occasions to return to this without thereby claiming to reconstitute the history of the one—that would be the task of an immense treatise. This history is in sum *one* from Heraclitus to Derrida and to us, but from the start it has displaced, uprooted, and overflowed its own unity as well as the "one" for which it apparently had taken responsibility.

(To avoid talking further about this now, suffice it to recall for a brief moment a few flashes or lightning bolts of Heraclitus's one or those of Aristotle or Plotinus, the unity of the Augustinian Trinity and that of the God of Averroës or Thomas Aquinas, not to mention that of Ibn 'Arabī or Eckhart, the monad of Leibniz, Hegel's One-for-itself, or Stirner's unique. One knows that these few comments are small in comparison with the immense and prolific course of this history. Perhaps one might be able to characterize this history someday as the history of the One, the unique history of the One, its advent [*avènement*] and its incarnations[4]—"one day" when the One will be disunited from itself. In fact we are already there. This is precisely what we are discussing.)

After Derrida wrote the text that I cited above, "more than one" became a favorite expression of his—I could even say, a fetish expression. More than one language, more than one session, more than one law, and so on, and perhaps more than one "one"—whether he risked this trick or not, it has been prescribed into law. Prescribed by the logic of the original supplement, to reuse his terms, but prescribed into this logic itself since the oldest point of departure (was there one, only one?) of philosophy.

More than one philosophy? More than one philosophy itself? The unity of philosophy has never ceased to be a point of contention within philosophy itself. Sometimes there must be essentially several philosophies that diverge or confront one another. Sometimes there must be a philosophy that continues and resumes itself unless instead it comes to announce its own "end" and its opening to another "thinking." Philosophy itself presupposes a distancing from a unique principle of the *sophia* of which it speaks. *Philein*, for its part, is a principle of nonunity: It implies the possibility of variations, distances, and approaches, and by principle then it holds at a distance the unity and the unicity that one would always obstinately want to attach to a *sophia*.

This also suggests that philosophy is always and constantly in one way or another *more than* philosophy. More than a philosophy and more than one philosophy. More than philosophy itself—this is what *philein* wishes to indicate—whereby philosophy would necessarily and originarily bring into play an excess onto unity: onto its own unity as well as the unity of its theme (whether it is called "being" or whatever one wishes; here, precisely only the excess of all signification or significance onto *one* sense counts in the end, whatever it is). "One way only" or "unique sense" [*sens unique*][5]—this is also one of the major terms for our historical process and its indictment.

"More than one": evidently this wishes to indicate more than a thing. (And what is more than "a thing" in this particularly indeterminate sense? A signified, a referent, an object, a concept? In truth this simply wishes to suggest "one," any unity, the fact that the intention of "meaning-to-say" [*vouloir dire*] may be filled by something other than "nothing"—thus simply that "one" is not nothing, and that for this reason the threshold question of the "one" is the question of nihilism: nothing, or the one).

"More than one," above all, suggests "more than one alone," more numerous—this expresses number itself in this sense, or numeration. More than one: one, two, three, four . . . (in nonrhythmic succession). This immediately implies a plurality of several "one"[6]: the one-one is followed by a second one, then another. Yet at the same time the plurality of several "one" opens the question of its nature: Is it an addition, a multiplication, or rather a distinction, a dissimilitude?

If the plurality is that of addition or multiplication, "more than one" could be extended indefinitely as the series of numbers, as the series of all possible numbers ("natural," "real," "imaginary," "irrational" numbers, etc.). The principle is that of numeration and numerousness: more than one, that is, not only a few ones, but many. Or more precisely, there are never a "few ones" without there being "many" on the horizon. Many, the multitude—the multiplication of the ones—which are not placed back under the jurisdiction of a One. This is quite simply because there is no jurisdiction nor any capital "One," but only the enumeration of the "one" and "one" and "one," and so forth. It is the enumeration that is the principle, the crowd, the numerousness whose number does not cease to grow. Addition is what always places the indefinite sum—which will never itself make a unity—farther away.

When humanity reaches the count of seven billion individuals—nine billion in forty years as projections predict—this large number seems to disseminate within itself any supposed unity of "man," of the "human genus." In fact, the human genus proves itself to be more like the genus of an indefinite multiplication than that of the growth of any one generic unity. It could thus prove itself to be the genus of no genus, like a blending of genera, or more precisely, a blending of the species because *homo sapiens* is a species of the genus *homo*. It is a species that does not let itself be grasped as such—that is, it does not let itself be *specified* or grasped under one *aspect* that would be proper to it: the *aspect* that in sum is a term akin to "species." The species without aspect, or the species whose characteristics would be reduced to what defines in principle the zoological "species," namely, the ability to interbreed. The human species would not develop any aspect other than its own exponential reproduction accompanied by a reduction or a destruction of many other living species—and a genetic transformation of even more, including itself, in processes such as cloning. More than one, yes, up to nine billion, and not *one* man? Can we imagine this: that a specific identity deploys itself as the pure multiplication of units that have a tendency to have value only as counting units in the counting that is itself endless? The seme or seed of "man" disseminated in its pure dispersion?

At the same time, multiplication makes apparent the other value of the "more than one." Dissemination—this other word from Derrida—takes away sense (and the seme and the seed) from unity and unicity. "More than one" is multitude—less as proliferation than as efflorescence, overabundance, and finally as excess of sense.

Without a doubt, the growth of the human population itself follows from the activity of man, who never ceases to pursue anything other than the reproduction of its conditions of life and that of its species. This constantly paves the way for unforeseen lives—lives not inscribed in the table of species and that exceed by far all that mutations are able to introduce in terms of novelty among the living. As more than one, and more than a species, man is itself a mutation of the living: it transforms life as maintenance into an enterprise (if one is permitted to retain the initial sense of the word, which speaks of nothing more than "enacting").

As long as life is maintained, it follows rhythms, alternations between sleep and wakefulness, action and passion, spoken exchanges, languages, and peoples. When life transforms maintenance into an enterprise, it undoes rhythms—indeed, it creates new ones. It invents complex cadences and yet it also brings any cadence into an indefinite acceleration. This is another aspect of the passage into great numbers: The population, the pace, the dimensions of financial bubbles, or the measurements of the universe, all fall under the sign of large numbers. Large numbers receive and produce at the same time this movement that one calls "mondialization" [world-creating] or "globalization," which, under one name or the other, suggests the presumption of a unity. The "global" unity, however, cannot be anything other than numeric, numerous, cumulative, and dissociative at the same time. It is the "more than one" [*plus d'un*] unfolded into a "one plus one plus one . . ."—in which case the "plus" [*plus*], which is ambiguous,[7] is used as an addition sign rather than as an indication of overcoming.

In addition, the unity of the whole is not distinct from the uniformity of the operation of adding. Nevertheless the value of the "plus" completely changes between "one plus one" [*un plus un*] and "more than one" [*plus d'un*]. The use of the former can be traced back to an arithmetic writing (+) and our use of language reserves another distinct pronunciation for it because the *s* must be pronounced once it is not used for the other values of the word.

In the use of the latter—the "more than one" [*plus d'un*]—the entire point is to get past the unity of the one. The point is not only to be able to count several "one" or several times the same "one." What is at stake is neither several nor many, but the status of the "one" itself. This does not uniformly preclude addition and multiplication: Derrida's text precisely says "the double did not *only* add itself to the simple." Addition is not excluded; it is traversed by another movement: In passing from one to two and thus from one "one" to two "one," I am not *only* juxtaposing unities—which, moreover, are necessarily supposed to be identical inasmuch as one can be added to the other. I am also affecting the nature or state of the "one" (and by consequence not only the nature or state of the first "one," but of the second and any other "one" to come).

This affecting is both a division and a supplementation or a supplanting: there is indeed one given as one, which is then followed by a second one, and yet because of this fact the first "one" cannot remain motionless in itself. On the one hand, the first "one" has itself passed into the second—precisely in its position of being a "one"—and because of this it is divided from itself. On the other hand, since this division removes from the "one" its initial property of unicity—and because we have truly not yet begun in this manner, not yet truly counted "one plus one"—the beginning or origin discloses its property *in place of* what should or could have begun. We tend to represent this as that which would form the "one" of the first step [*pas*], even though a "step" in and of itself divides itself.

One could say that there are two logics of the "one": one that posits it as motionless but raises the difficulty of knowing how to *step* into [passer *à*] more than one, and one that receives it as the *step* [pas] itself, which divides itself and supplants and supplements its initiality on its own.

"One" divided is not "one + one," but rather two "one" stemming from the division of a first that did not take place. Two times, then, plus the division itself—as the initiality of what has not "begun"—and hence a third time: a ternary, the metric figure of which could be the meter called "amphibrach" (˘ ¯ ˘), although this does not exclude other possible beats, nor does it exclude the possibility of passing to four by splitting into two the second time. As we know, this is what occurred in the Hegelian dialectic as well as in the Christian Trinity, which can be considered as the unity of three with unity itself, which is called "God" (or rather, in the human analogy employed by Augustine, the unity of memory, intelligence, and love with the person who is herself nothing more than these three properties put together)—the fourth time does not indicate any additional "unity," but the extension or internal tension of the ternary. Thus, the *Geviert* of Heidegger is the figure of the extension or disclosure of the world. It is the extension within which existence is possible according to *Ereignis*, which can be understood as the event of the one (of each one, and/or of the one of "Being") as well as the division or partition of the one.[8]

II

> This being [the "one"] must become other for himself, reject himself, condemn himself, abolish himself, to the profit of the others in order to be reconstituted in their unity with him.
>
> —ANDRÉ BRETON, *Communicating Vessels*[9]

I am not setting out, of course, to skip through the figures and concepts of three and four in philosophy, theology, mysticism, and alchemy. I only want to recall how important this numeration has been throughout our history. From the outset, and long before philosophy, an entire numerology and arithmosophy has integrated or doubled enumeration. The "one, one . . ." has always already given rise to speculation about the "one" itself—the "one" that is not enumerated. Nevertheless, beyond any numerology the issue is to *hear* [entendre] because there is a beat, a beating—there is a pulsation.

Perhaps "one" is never possible without this pulsation. This pulsation is not numeration even though it is already at play within numeration, which already beats out the simple "one + one + one . . ." There is no addition without repetition, and no succession without meter. No chanting when reciting the addition table—"one plus one, two; two plus one, three . . ."—without a hint of song, of incantation. It is not then the counting that counts. It is the resonance: it is the language of calculation referring back to itself that which exceeds both calculation and language.

"One" perhaps would not have to do with counting but with rhythm, not with the number of the downbeat—3, 4, and so on—but with movement, tempo, vitality, and pulsation. For a long time there has been a dispute—particularly between the Platonic, Aristotelian, and Neoplatonic traditions—about whether one or being is primary. Pico della Mirandola wished to bring this debate to an end when he wrote his *De Ente et Uno* [*On Being and the One*]. His solution consisted in overcoming the rivalry between the two candidates by attributing both (along with the good and truth) to God, who does not allow himself to be grasped by thought.[10] Because God's unity is "prior" to that of the one itself, it is actual infinity at the same time as it is intimacy unto itself insofar as its "true knowledge" is a "total ignorance."[11]

In other words, one of these things is not like the other one. There is more than one buried in the heart of the one—and also in the heart of being. Although, on the one hand, it is, in effect, impossible for a being to be without being one, it is possible—perhaps even necessary—that the one, like being, belongs to a different logic of the One. By moving away from the substantive "being"—which, in effect, is subject to numeration from the outset—and toward the verb "to be," one comes to think about this verb as a transitive verb (Heidegger asks this of us sometimes). The verb "to be" "is not [*pas*]," as Heidegger says, and for this reason it beckons to be crossed out or put under erasure. This erasure is not a canceling out but a spacing [*écartement*] (a quartering out [*écartèlement*] if one wants, or a fissure in the shape of a star [*étoilement*]). This ungrammatical transitivity could be rendered thus: to be splits and spaces [*écarter*] being.

To be, then, is not "one" being, nor is it the production of beings one by one. Rather "to be" forms the act of spacing, the act of distinguishing between and dividing beings. This act of spacing is one—without being "one being." "To be" is the unity of acting—its transitive unity, which does not precede the "ones" of beings and does not occur to them either, but is, or rather, acts, and takes note of their event or advent.

This "one" does not allow itself to be counted. It is the one of the push, the pulsation, or the pulsion or drive [*pulsion*] that allows what is to be. It is a rhythmical one—that is, both a composite one that is plural in itself and a one that is subtracted from the count. Here one does not speak of counting the measure of the rhythm. One speaks rather of clapping the rhythm in one's hands or beating it on a drum, which is something else altogether. One speaks of dance and tempo. The student of dancing counts his or her steps, but not the dancer, who dances them or is carried and propelled by them.

"To be" is to be pushed outside, that is to say, to ex-ist. Each existent is one existent. This does not mean that each is *one* in itself and for itself, unique and united. The pulsation that projects each and every one belongs to quite a different kind of unity. In this sense, the world is indeed "one." Yet this unity lies hidden from us because of its vast number, not because of its sheer amount—this greatness itself seems to bask in adding for the sake of adding itself—but by the fact that beneath numeration and multitude we struggle to understand what kind of unity could reveal itself.

Without a doubt this pulsation does not reveal itself in the sense that we have given to the term reveal. It does not come into the light of day. It does not let itself be known or counted. Rather it shows itself always and everywhere. In effect, always and everywhere in the extravagant profusion of nature as well as in the proliferation—the swarming even—of our machines, our signals, and our ends without finality. Always and everywhere there appears to be continually more of something like "one" sending off [*envoi*] that is not projected but thrown, ejected from the world toward an outside that is always more vast and further away from the assurance of a unity of design, aim, or history.

It might be in effect that the One prior to any unity escapes not only our grasp—which it has always done as we have always known—but withdraws itself from itself. Another way to say this: "God is dead."

Still the other side of this withdrawal could be that this prior One, the eminent one or "*maxime unum*" as Thomas Aquinas puts it,[12] withdraws itself from all unity and numeration. By withdrawing, and in effect leaving the deserted corpse of the "one," the One indicates a higher, more immeasurable, more uncountable unity—and thus one that is also an innumerable one and carries the innumerable of large numbers. And yet this unity is "one," in that it exceeds every multiplicity, every diversity, *and hence thereby* exceeds every countable unity in a multiplicity.

One from beyond or from before the One. It is the punctual One without dimension. It is that which has not taken place—neither place nor time. Rather, it opens all possibility for time and place. This excess has nothing magical or mystical about it. It is not a foggy, metaphysical entity. Excess is given within the nature of the "one." If "being and the one convert into one another," as the classical doctrine of the transcendentalists suggested, and if "to be" is to be pushed outside (outside, in front of, farther away, elsewhere, however one wishes), then "one" gives the energy of this push as a rhythmic pulsation.

The "*maxime unum*"—the greatest one that is—is greater in that it is "*maxime indivisum*," as Thomas Aquinas also says. To be supremely undivided is to be unsusceptible to division in any manner: not into parts, or substance and accidents, or activity or potentiality, or finally into "any thing" or the nature, position, or existence of that thing. The exceeding one, by exceeding itself, withdraws itself from any assignation of this kind and can only be considered as this unity and unicity of pulsation, pulsion, impulse. It is not a "one." It is only the absolute simplicity of a push, a leap, an impetus.

Its unicity is absolute, and this is why the "*maxime unum*" is indeed also "one and all" according to the unity of all attributed to Heraclitus, who gives us the saying, "The one conjoins with itself by differing from itself." The exceeding one "is" very precisely by differing from and by deferring itself in the push that opens, spaces, and disperses all things. Still it does not differentiate on its own nor does it resolve itself in differences. It differs from and defers itself—which must be understood according to the logic of *différance*: This differing and deferring does not occur to it but is itself this differing and deferring. If one may say so, it "is" only—but integrally—the movement. It is the push, the pulsion that carries itself and that carries forward—one that communicates itself to all existence and to all existences. One must not overlook the fact that in the famous "one differing from itself" of Heraclitus, the verb *diapherō* must also or first be understood according to its value of transporting, movement, and trembling that carries one away.

The exceeding one is a carrying away. It is an impulse that follows from the very fact that there is a world, or worlds, as many as one would like—that there is some thing. It is the secret one of this indeterminate, anonymous, and discreet "something," the discretion of which bursts forth at the same time into indefinite multiplicities of unities and unicities. Thus one could say, following the logic of *différance*, that this carrying away does not carry away anything that would have preexisted it, not even itself.[13]

This one is another "more than one." It is—inasmuch as it "is"—the "more than one" that transforms itself into "more than only one." More than "one," but not more numerous—on the contrary, more one than one. Not counting but transporting. Not unity but sending off [*envoi*]. Not given but giving. Unique, but not isolated—on the contrary, taken up again and thrown into every meter of existence, each of which is unique. One of absolute unicity and not of union: It does not reunite anything within itself since it opens the *pas*—the passage, the bandwidth [*la bande passante*]—of the diverse. The *pas*—which is itself divided—of the division of all things.

As Blanchot writes: "The 'One' is what least authorizes union, even with the infinitely distant; still less does it authorize mystical elevation and fusion."[14] Not confusion, but profusion. The one that is not added—or "not only" added. The one that transforms the very sense of "adding," and that is not reunited in itself. The one that is neither additive nor inclusive is not pulled back into itself in the way that a point is (for example, as the Kantian

"I" is). On the contrary, it is the transporting of itself and out of itself (though it has neither outside nor inside)—a transporting of the point, in a sense, but a transporting that creates a line. It is not a continuous line but a rhythmic one. It is the cadence of the always renewed plurality of all singulars. Each singular is itself "one" in this sense: in that it repeats the initial transporting—a transporting of the initial—as it is itself a prelude to other profusions (a life, for example, all its moments, all the bumps along the way, and its death).

If there is a unity of the world—or more exactly if "the world" can be thought of as the nonnumerable unity of Being, of this being understood as a rhythmic singularity according to this "ternary essence" of the infinite that Derrida talked about—it must be a unity of this order. In other words, the order of the profusion of the species, aspects, and manners of the existing, of the fertile push of the living; and also, another proliferation—that is, the proliferation of our machines, devices, instruments, and motors, and of our transplanted, transorganic, virtual, fiduciary, and fictive organs.

Yet this requires that the proliferation and profusion must be an abundance and not an accumulation by addition. It is neither the addition of one moment of time to another, nor the addition of one portion of space to another, nor one individual to another. We know that the most abstract general form of this addition is currency or money: the equivalence and exchangeability of addable units (of *numéraires*, as we used to say).[15] We also know that it is the exponential development of financial techniques (credit, stocks, banks, etc.) that has allowed the deployment of other techniques, and with them our entrance into the era of large numbers—cosmic, demographic, and electronic. We do not know whether these numbers are those of swarming, excess, and congestion or rather of abundance, generosity, and wealth—still, in what sense of wealth?

Wealth—this can be another word for "more than one." It shows us very clearly the crossroads between the accumulation of countable units and a generosity that can come only from elsewhere and before, from the unity of an infinite rhythm. This is perhaps precisely why our civilization has been so preoccupied with the one since the beginning: because it knows that it lies between the two, between the one that adds to itself and the one that carries itself away from itself.

Between the one and the other. Today we know that we have to rework all of the terms of this alternative—or rather, of this interval, contradiction, or complication. Between the one and the other one. This implies at once what is in between the one and the other. Is the other the other one or the other of the one? And of which one? To be continued.

J.-L. Nancy

Less Than One, Then

I

The one as a pure singular, an ontic unicity, a primary principle, a decreed edict, an identity unto itself, or an incorruptible [*imprescriptible*] *archē* or timeless principle has therefore not taken place. It has never even begun to take place. It must be that the knell [*glas*] rings for the one's fantasy (its phantasm, its phantom or ghost). Or rather: The echo of the mallot on the cast iron must be heard as merging with its own repetition, so as to place oneself in the stretch of time in-between the strikes that ring the bell. This stretch of time or interval is not a fading into silence nor does it announce a new toll. It has no sense in itself; it is not even a sign. It is precisely the step [*pas*], the exiting, the outside coming into being that forms the already fleeting sound and opens the horizon of its potential and always uncertain repetition. It is a mediation more than a meditation. It is an impetus or energy—almost an intrusion. The knell, however, did not use to announce only death as we sometimes believe, but also the throes of death, the disappearance that *will* occur, the end as it is taking place, happening, or in process. The knell is a dynamic rather than an ending. It passes by and becomes a guide [*passeur*]—not from one world to the other but from *this* world to *itself.* Rhythm, which structures each occurrence of sound—incessantly, obsessively, repeating only one and the same note—is precisely what begets or engenders the significant unity. The first reverberation of the knell is only designated or denoted when the next is heard ("The bell sounded a second time. 'It is the knell [*glas*], Monsieur Madeleine.' "): not one without two and not two without distance to the one.[1] This, effectively, makes three. Still can the incongruous be ordered into a sum?

Glas, of course, is also a margin, a gulf, an interstice—both an immense rift and a microscopic fissure between Hegel and Genet. Derrida's text does not presume to be exemplary, nor does it have the intention of establishing itself as an archetype.[2] On the contrary, *glas* is the singular—and yet rhythmically universal—construction of an originary incision. The interval between the two columns—which is often repeated within the very columns themselves—is precisely what allows each line, each voice, and each range to be constructed by differentiating itself from itself. The interval's relationship to the other is antidialectical. In other words, it is a manner of unscrewing the syllogistic cog in the wheel from within. A threefold essence to be exact. Texts interlacing without constriction. Flows and movements without equilibrium. Hegel/nothing/Genet. Nothing, which is obviously where everything comes into play (which is perhaps another name for Derrida), is the caesura that allows the disunion of the One and consequently its existence. Nothing is absence becoming tied together. This gaping void between the two columns of *Glas* is not a sketch that renders interpretation nor is it an outline of globalization. It is neither syncretism nor analyticity. It is the distance that gives to each thought and each measure[3]—approached as literally, precisely, and modestly as possible—the possibility for disenclosure.[4] Here once again the task assigned to the reader is threefold[5]: decipher one text, then the other, and then compare them beyond their nooks and crannies. Without the aid of any Rosetta stone, of course, to solve a hieroglyphic enigma that was skillfully constructed so as not to be resolved in any way.[6] Resolution—in the musical sense—is what would come to dissolve the tension or smooth out the dissonance. Resolution would thus be each note giving up on the One inasmuch as each note defers itself to itself by way of the others. This "deferring-to" [*différer-à*] is the reverse swing of the pendulum, and hence the putting into rhythm—and almost into resonance—of the "being-to" that supplants and supplements the "being-for" or the "being-by-way-of." Irresolution is imperative for sense to break through into a world that has disappeared or is disappearing.[7] Irresolution is this defining *margin*, this "footnote," which is completely de(con)textualized as it seeps into the body of the paragraph,[8] as it occasionally climbs up into the position of the title.

The cadence of the "More than one" is simple and unvarying: "two times, then, plus division itself as initiality." It is *almost* the rhythm of the hexameter, the sequence of the dactyl, or the abruptness of the phalanx. *Almost*, because in order to *first* hear these two identical times, the finger must be thought of as starting from what it touches or points to—starting from the world, in the finger's relationship to exteriority, and no longer as that which comes after the anteriority of the body that supports it. It is an inversion or uni-version [*un-version*].[9] Unless indeed we were to accept the temporal primacy of long duration—in other words, of division. The hexameter would then restore the initiality of division, the exiting, or the ex-tracting movement. Is it the influence of the proximal on the distal?

Clearly it is not insignificant that this meter is that of the epic poem or Greek myth starting with the great Hesiodic poem. Fittingly, in this poem Khaos is not only disorder, Kosmos's progenitor and complement (as its direction or horizon, as the reign of the Olympians *to come*). Khaos also plays the role of a gaping void, a primitive vacuity, a chasm, an abyss—in other words, here again, it plays the role of a division that initiates. It is the principle as much as the element, the conditions of possibility as much as the absence that has to be surmounted. Khaos, in all its ambiguity, is both the undifferentiated world and the very principle for its differentiation. It is the escaping, the "pushing," or the inchoative of the step [*pas*] of the directed movement, of the tension directed toward Eros or Gaia. It is the ruined.

The poet or bard is a craftsman. The hexameter is his material, his clay. If he thinks, naturally, that Apollo protects him, he is above all else inspired by the muses. The magical value of the poem proceeds from its *rhythm*,[10] whose artificiality reveals an immemorial origin, an origin that is foreign and thus universal. This artificiality already astonished Aristotle. This dactylic pulsation is also—and this is not a coincidence—the rhythm of the central melody in Schubert's quartet *Der Töd und das Mädchen*, or in the Allegretto of Beethoven's Seventh Symphony. Two protean and plurivocal works in which the composers explicitly wished to revive the preeminence of rhythm in order to depart from the possibility of any regulating structure. Music as a push that is incessantly reinitiated, as a uni-pulsive response, that is, music as the latency of all the modes (Fourier's analysis shows that the spectrum of impulsion is infinite) and also as a way of never allowing oneself to be satisfied with only one of the modes without the others.

The cadence or cadance of a ternary differ*a*nce has been deployed like a passage, a division, and a supplanting since even before what is traditionally called the birth of philosophy (besides one could argue that *philein* is likely unnecessary for the disunion of the One since sophism also—and maybe even more so because of its founding principles—unfolded by way of variations and distancing).[11] It permeates our tradition all the way down to those who criticize it, reject it, divert it, subvert it, or deconstruct it. Down to Nietzsche. Down to Derrida, down to Nancy. Down to Deleuze as well, whose rhythmical "line" becomes molecular in order to undo the molar unity, and whose multiplicity becomes an empirical operator.

Still the other sense of the "more than one," its numeric signification, its arithmetic and linear value, its countable and additive occurrence, is also not absent from contemporary practice. It has unfolded in another way. Antireductionism—and to a large degree antirepresentationalism—is precisely a plural and pluralizing gesture. Antireductionism may largely be found in the analytic school of thought: from Putman to Rorty, from Dewey to Davidson, from Kuhn to Goodman, and from the later Wittgenstein to Hook or Michaels. The analytic "more than one" is foremost that of enumeration: more than one possible world, according to Lewis; more than one constructed world, according to Goodman. The worlds of modal realism, created in order to give sense to counterfactual conditionals,[12] do not, of course, have anything to do with the worlds as correct and copresent versions of "radical relativism under rigorous restraints."[13] They have nothing to do with these worlds except that they swarm, they coexist, and they multiply. Influenced by William James's *A Pluralistic Universe*, Goodman shows that the distinction between monism and pluralism does not survive analysis: Worlds—or different versions, which amount to the same thing—are created by composition and decomposition, by balancing and assemblaging, by eliminating and supplementing, by analogies and distortions. This proliferation does not prove that our epistemic advances are in a transitory or incomplete state. It *is* the mode (all of the modes together, the meta-modality) whereby "relative reality" unfolds, whereby substance is dissolved into function and the given is recognized as conquered. Great numbers, sequences, sums, series, additions, accumulations, assemblages, conglomerates, heaps, groups, repetitions, and cumulations.

Must one conclude then that analytic philosophy has appropriated the literal, countable, numeric, or numerical "more than one," whereas the continental approach has been concerned with literary, rhythmic, different, and perpetually differed division? Nothing could be—at least systematically or normatively—less certain. First, the "passion of the one" is always and already lurking: For example, in the analytic field one must remember the unique, correct description to which, according to Williams, all attempts at denotation must be *reduced*, or the precedence—well beyond a simple methodology—that Fodor gives to the scientific approach consistently. Second, these modes of the multiple are not absolutely heterogeneous to one another and would collapse, in their very dynamic, an exclusively dichotomous and split taxonomy. Finally, in a more fundamental way, an inevitable porosity between these two different ways of deconstructing the one, the unitary, and the singular would have to be structurally established. And it is at the heart of the numerousness, the profusion, the abundance, the multiplicity, and especially the density of multiple worlds that this osmotic passage becomes possible and even inevitable. It is here specifically that addition can be transformed into difference. Here the two senses of the "more than one" *touch*. If there "never are a 'few ones' without there being 'many' on the horizon," there never are many—and the generating principle of many—without infinity crossing over the horizon. Tearing it, writes Deleuze; perforating or puncturing it, writes Derrida. It is the *limit* (in the mathematical sense, which indeed de-limits and dis-closes) that allows the simple sum to become an authentic extraction. It is an opening, without rupture or force, which belongs to another register. A finite sum of rational numbers is rational, but an infinite sum of rational numbers may be irrational, even transcendental.[14] The *limits* of sequences of rational numbers delimit or define real numbers. Passing from positive integers, which are defined by their simple numbering capacity, to abstract objects, to the fundamentals of real analysis, is achieved by passing through the limit. It is not a change in quantity; it is the creation—through boundless and precisely numberless quantities—of another substance, entity, or order, which is incommensurable to the substance, entity, or order from which it stems. It is the invention of the *continuum*. It is contact between *the* ones. It is a strange unisignificance.

Goodman's multiple worlds—selected according to their *correction* and *adjustment* (his relativism is the opposite of an ontic nihilism and an axiological laxity)[15]—strictly follow a numeric schema: These worlds allow themselves to be identified, located, and counted. It is, if one may say so, their "proper mode." Still they also generate their own principle of infinity: their propagation cannot be circumscribed by anything. Their possible variants, variations, and varieties are, by law if not in fact, unlimited. Thus they require one to think this passage *through the limit*. At the edge of these irreducible versions, at the fringes of these worlds that do not canonically transform into each other anymore, is a real that is stretched, spread apart, and quartered (that is, dismembered and hence dis-unified)[16] between these "ways of describing" that are reinterpreted as authentic, demiurgic acts. "How to do things with words . . ." once again, once *more*, but in quite a different sense than Austin gave to the phrase.[17] Authentic *and* artifacted worlds—worlds that are authentic because they are artifacts. The Goodmanian mechanism became a hybrid in the margin, inevitably allowing the passage from one "more than one" to the other, creating a diverged convergence.

On the other hand, there is clearly no doubt that through its distinction and division the One inexorably brings about numeric plurality, as it is the principle of "supplanting its initiality" that leads to multiplication.

Thus what is indicated here through the constructivist example is that the crosstie or railroad tie can be used in an unexpected direction (that in the end the breach opens in *both directions*): the one + one + . . . allows for the other unity's condition of pulsation, which is also the other of unity. The limit is the mode of this transmutation.

Putnam thought that the Derridian line of argument had its source in Saussure and Goodman, borrowing from the former the concept of incommensurability and from the latter a radically unrealistic system.[18] Using the "more than one" as a measure of comparison, the gestures of deconstructionists and relativists are not easily reduced, analyzed, or synthesized. These gestures present themselves rather as openings, bridges, and rifts that allow the transition from one multiple to the other. At this point, one could try to rely on Goodman to cleanse Derrida of the nihilism for which he was once reproached, and to depend upon différ*a*nce in order to breathe life into the primary, mysterious dynamic that is indispensable for "making worlds." Goodman would bring a rigorous nominalism and a symbolic theory—and hence the necessary conditions—and Derrida would constitute the acting out, the *praxis*, or the operational exercise. This would be simple, efficient, and in all likelihood, just, but it would obviously still be insufficient. As Deleuze and Guattari never ceased to remind us, to look for a system's shortcomings, imperfections, or inconsistencies through the lens of an other ([de]con)struction is always to do violence to the system—even if it were, like the *rhizome*, a revocation of systematic thought. The "more than one" is indeed the nonplace of a possible meeting between "French Theory" and analytic philosophy (two referents without referees, certainly; still, what is really in question here are the borders, edges, and lineaments *themselves*)—under the strict condition that the proper modes of a particular recomposition of the structurality of discourse are put into *operation*. The Deleuzian repetition as an accepted instability,[19] the Derridean proof—re*iterated*—of the inconsistency of an opposition between sensible (hence degraded) and intelligible (hence original) repetition,[20] the Goodmanian repetition as a process of transformation,[21] the repetition that is freed from all re-presentation according to Rorty[22]—all of these are possible junctions, passages, or links between the two possible modes of extracting the One, between the two ways of overcoming an ossified or stratified monomial. This is what Searle and Austin, for example, never dared to do even though their approaches led them to it almost inevitably: to accept the nature of discourse as being universally made up of quotations[23]—its archi-scriptural dimension—as a means of auto-differ*a*nciation, as a possibility of an internal pulsion or drive. On the contrary, they saw in this the symptom of an undesirable sensitivity to context that would have to be elucidated, eluded, and even, *in fine*, eradicated.

The two modes of the "more than one," whose senses—and perhaps essences—differ without nevertheless standing in opposition or bifurcating each other, are thus a point of contact between currents of thought that are supposedly dis-jointed and autonomous. As with any point of contact, this one is also a point of both traction and friction. A space of zero measure in which the disorder inherent to the différ*a*ntial movement inevitably spreads in directions that are independent from the initial pulsions. It is not so much a "common vibration," which would inevitably be tarnished by what Salanskis calls the impossible injection of the "motif of iteration in the analytic problematic,"[24] as it is different modes of existence—in the precise sense that Souriau gives to this expression[25]—of complimentary divisions opened up by the one for the other, by the *One* for the other*s*. It was perhaps not unforeseeable that analytic and continental thought, in both their oral and written claims, would conjoin around the common need to revoke a certain "order of the one." Surely we find each other again precisely where we left each other . . . but much more significant is the new assemblage that can result from it. It may not be just a mere point of tangency: Something here seems to profoundly intermix these supposedly antagonistic approaches unto confusion and profusion. More than a loose affiliation and less than a convergence, a sort of in-between-existence is on the way perhaps. An in-between-existence without dis-affiliation.

Goodman, Derrida, and art. *Archē*types. Derrida asks the infinitely ambiguous, polysemantic, and recursive question of the "idiom in painting" from the start.[26] Goodman immediately introduces the subversive and unfounding [*défondatrice*] enigma of the "languages of art," be they nonverbal or nonliteral.[27] The same obsession of difference or bifurcation (the "Divided Prime Mover"[28] of Derrida, the work of art and its forged copy for Goodman).[29]

The same interest for the frame, mounting, title, and signature (in the words of Derrida), for the symptoms, conditions, domains, and limits (in the words of Goodman). This is not at all to suggest that Derrida and Goodman claim, write, or imply the *same* thing. That would be at best paradoxical and at worst senseless for these thinkers of the untranslatable. Nothing is ever the *same*. Especially not in this instance. It is rather a question of comprehending and articulating them and putting them into r*ea*sonance [*r*ai*sonance*] in a dynamic of the "more than one," in which case they would form the initial repetition of this dynamic. One, then the other (Goodman, then Derrida)—at one and the same time touch tangentially, brushing or grazing past one another *and yet still* arbitrarily distant—and then the rhythm of their reformulations. Two consciousnesses, which are stopped and started in their mutual rewritings. Evidently, it is of little concern whether they know and acknowledge one another or not.

II

Although "More than one" is thus the pre-text for a mutual rediscovery or sharing of phenomenological and analytical explorations, its meticulous development is not only limited to philosophy.

It is clearly not by chance that "more than one muse" exists, that the world is "dislocated into plural worlds, or more precisely, into the irreducible plurality *of* the unity 'world': this is the *a priori* and the transcendental of art."[30] It is also not insignificant that the attempt to subsume fine arts under a unitary concept of Art is essentially concurrent with the tendency of artistic practices to become radically autonomous. There is a strange tension between an implied unity of purpose and the proliferation of manners, methods, and practices. Whereas Batteau writes "art" in the singular, thus leaving an indelible mark on Diderot, the Schlegel brothers, Kant, and Hegel, Lessing departs definitively from the *ut pictura poesis* by demonstrating that the agape mouth of Laocoon obeys an independent logic of its own. Take the hierarchy of Horace, whose friend was Virgil and whose benefactor was Maecenas. This hierarchy, which no longer has any significance, was debated throughout history and overthrown more than once up until the

Renaissance—to establish it definitively was the goal of Leonardo da Vinci's *Treatise on Painting*. It is remarkable that it is precisely in the context of a thought of convergence, unity, and to a certain degree, the sacredness of artistic inclinations and interests in which this very dynamic unfolds, a dynamic that acknowledges individual specificities and hence different ways of putting art into *works* [*mises en* œuvres]. Adorno theorized about this in detail in relation to modern art,[31] but the origin of it is clearly anterior. At the very heart of the desire to conceive—or to invent—the One of Art, a difference in methods and approaches reveals itself. There is not only a diversity of methods but also an inner, true fissure, an internal push or rupture. Harmony, writes Jankélévitch, is "less the rational synthesis of opposites than the irrational symbiosis of the heterogeneous."[32] This regime of symbiotic heterogeneity is precisely a coexistence in which the host and the parasite-partner get rid of their individual identities without completely merging.

Paradoxically, the "ternary essence," the essence of the one and infinity, is also the dynamic within music as well as the tension that is extrinsic to it. Paradoxically, because what is at issue here is not to refer to ternary *time*—the time of jazz or New Orleans, of swing or boogie—but rather to think about music in its spatial or spectral sense.[33]

The Goldberg Variations is a doubly "unitary" work. First, because contrary to the far too rigid, almost crushing instructions that they were subjected to at the time of their reverent canonization, before this time the variations could be played and listened to independently of one another. "Play me *one* of my variations," the old Keyserlingk used to demand according to the legend of Forkel. Additionally, because like the *Musical Offering*, the *Canonic Variations*, and *The Art of Fugue*, *The Goldberg Variations* are constructed and organically structured around a unique motif. Rigorously following the logic of the Baroque *Fold* [Pli], all of the variations are already there, latent, virtual, in a state of potentiality—compossible—in the initial *Aria*, a peaceful melody full of restraint that is discreetly ornamented in the French Baroque style. Nevertheless: It is indeed the *three* that is in question here with regard to founding the One. Every three variations—on the third one, the sixth one, and the ninth one—the bass produces a canon at the unison, then at the second, at the third, and so on. The ninth canon—cosmic order, that is, the twenty-seventh variation—carries out the divine number (3×3×3) and reiterates the message that was announced a few years prior in the third part of the *Clavier-Übung*.[34] All of the canons are evidently for three voices (except the ninth, which is for two voices: three to the power of two is nine). The principle here is more trinitarian than ternary. One could endlessly develop, as has been the case on several occasions, a mystical, magical, and esoteric interpretation of the arithmological lines that are interlaced and interwoven by Bach. Yet this is not what matters here: the nodal point is the *identitas in varietate*. A systematic exploration of all the possible variations on an extraordinarily naked theme, a creation of unheard-of worlds from a material that is intentionally outdated; a syncretic speech, obsessively coherent and skillfully ordered down to the imperceptibly subversive smile of the final *quodlibet* (doubtlessly the same smile that can be discerned on Bach's lips in Haussmann's portrait of him; the *quodlibet* with three enigmatic notes, with *three* different lines or tunes announcing a *triple* canon). All of this is there and much more. More important, with this *addition* of the thirty variations that altogether supplement and ornament the initial *Aria*, there is the invention of a true *Alter Ego*. The *Aria Da Capo*, which is (re-)heard following the pyramid structure that shapes the *Overture* of Variation 15, has nothing to do with the initial melody. And yet it is perfectly identical to it. Number, numericality, abundance, variations, many variations (evidently more than Scarlatti—as many as the master Buxtehude) were needed in order for the initial one to reinitiate itself, that is, for it to be both perceived as what it has always been—reanimated—and radically revised, renewed, and reinvented.[35] An eternal return (that of Deleuze *à la* Nietzsche, a departure from nihilism, an ontological displacement).[36] An eternal delay. An addition and an overcoming: plurals so that the primitive singularity may derive its sense from a push that extracts this singularity from its staticity and its autarchy. Thirty variations like an interval between two sarabands, like a universe between the work's theme and its reprise, like a mediation between the one and its splitting into two (its repetition, its belated arrival). The first *Aria* is fundamentally affected by the existence of the second, by the immeasurable distance (be it a spatial or temporal measure, be it even a measure in its rhythmic meaning, or even the superior unity, that is, the *carrure* or balanced symmetry that separates and conjoins the arias: it is the variations—as a step [*pas*], exit, or discord—that have unified *and* disunified the arias. Here too it is a passage or confusion from the substantive to the transitive: The substance *is* the transition. A counterpoint transition (more than a path [*voie*], more than a voice [*voix*]) in thirty canons, fugues, gigues, and chorales that are and that form the *sense* of the motif of a bass that does not have a beginning.

There is no exposition and reprise. And there is certainly no theme and variations. There is birth and conformation *through* variety. The *Aria* does not preexist its reprise. Instead, it is through these that it makes sound. It is not what the work invents itself from; it is both the retrospective and projective introduction. It is a musical Ouroboros in which both the final and initial *Aria* join and unite in the infinity of their spacing in order to give body to this strange assemblage of rhythms and notes. Without a head or tail? It is unified by its split in two.

The relationship between music and theater is quite ambiguous. The song of the chorus is the bedrock of tragedy. It is both its foundation and inner tension. Following Schopenhauer, Nietzsche liked to envision the birth of dramatic art as a Dionysian pulsion, a Bacchic procession, a climactic dithyramb, a pagan form of dance that was both opposed to *and* ordered by the harmonizing power of the chorus. A unique, united chorus—in unison—which is also the expression of the fundamental and irreducible division between what has been and what might have been. It reminds, it hammers home, it proclaims that a choice is never without cost, that the downside of decisions, their dark and sometimes hideous side, is as fundamental as the legitimacy bestowed upon them by the urgency of circumstances or by what was of interest to the city. Inevitably, Agamemnon had to sacrifice Iphigenia. But the chorus relentlessly reproaches him for not having taken responsibility for the duality of his status as both king *and* father. The chorus is a manner of instating a primary dyadism (which is itself divided and therefore ternary), of making the past facts and the possible facts copresent, coactual, and cotrue. It is the *movement* of the real's realization *by* the thought of the virtual. What is given again is the rhythm: the tragic event, its repetition (its reactualization, its echo) and the chorus as division. When he speaks of the "Theater of Cruelty," Derrida writes that the dialectical horizon is "the indefinite movement of finitude, of the unity of life and death, of difference, of original repetition, that is, of the origin of tragedy as the absence of a simple origin. . . . [Primitive theater and cruelty] thus also begin[s] by repetition."[37] The dialectic that is in question here is evidently not to be understood in the sense of a "conventional Hegelianism": It is, strictly speaking, the antidote to the idea of a "pure origin" or "spirit of beginnings." Tragedy, by raising what should or might have happened to the level of the ontic authenticity of recognizable facts, therefore inscribes the deferred—and eventually latent—repetition onto the very heart of theatrical gesture. This gesture is originarily ternary: it is structured as much by what it lacks as by what it exhibits. In other words, the act, the act that is lacking [*acte manqué* or parapraxis, as in a slip of the tongue] and the distance that separates the two, which is established and underscored by the chorus, form the primary disunion. It is all the more striking that this argument presents itself in the midst of a reflection on Artaud, who indeed connects—or subordinates—the birth of theater to that of *dance*. (The manner in which Derrida manages to bypass and then reverse and invert the method of Artaud but without distorting it is very surprising and subtle. Focused entirely on rehabilitating the production of a work to the detriment, if one may say, of textuality or literality,[38] Artaud's approach is replaced by, reinserted in—almost reintegrated into—a movement that deconstructs logocentrism *via* the rule or constraint of de-theologizing the classical tradition).

What we have proposed regarding the *Goldberg Variations* could also be written about the *Diabelli Variations* or the *Paganini Variations*. And, beyond Bach, Beethoven, and Brahms, about all music, which is intrinsically variating or variational. The example of the *Goldberg Variations* is paradigmatic. It makes manifest, through its structure, the role and presence of the "more than one," but it does not present any specificity or any principial particularity. Likewise, what is true for theater is also true for poetry and literature. (*I Bach you, yes I Bach you*, repeated Ghérasim Luca, thus inventing stuttered sense). This must be shown in detail, and the task would be immense, but one can surmise it without much risk. The examples abound, and if they do not become the rule or law it is simply because a rule without exception is evidently impossible in this domain. Is putting this plurality into light—or into chiaroscuro—not what is at stake within a large part of the syntheses and analyses enacted by Derrida of the work of Nietzsche, Blanchot, Joyce, Celan, Jabès, and many others?

There is nevertheless a domain in which the mechanism of the "More than one" seems to get jammed: that of the sciences and in particular the hard science *par excellence*, the science whose pores seem to be closed off definitively to plurality, the most mathematical yet heteronomous of thoughts: physics.

There is no *philein* to temper, disseminate, or reject the kind of knowledge that belongs to physics. This invincible citadel of objective or positive knowledge, this temple of the actual and the factual, is unitary in its essence. This fact is so evident that to question it would seemingly lead one to insult the entire approach of physics. It is sufficient to simply observe that a conceptual and normative unification is at work in Kepler, Newton, Maxwell, Glashow, Salam, Weinberg, and others. This unification is at the heart of an extraordinarily coherent edifice. And this edifice—it is apparently indisputable—has managed to subsume the chimeric diversity of the real under a reduced number—which is inescapably aimed toward unity—of principles. Multiples are artifices or artifacts: the plural is inevitably disintegrated for the sake of a more accurate observation or a model of superior quality. It is in a transitory and intransitive state.

And yet something seems to evade this well-wrought schema.

In the heavens of the contemporary astronomer (who is more a chaologist than a cosmologist), comets run alongside planets, planets run alongside stars, stars run alongside white dwarfs, pulsars and black holes, and black holes lurk in the heart of active galaxies including Seyfert, *Starburst*, quasar, blazar, BL Lacertae, OVV, and radio galaxies, and many others. Intense, at times relativist shocks unfold over considerable distances following complex magnetohydrodynamic processes. Interstellar and intergalactic environments develop a subtle chemistry. Accretions and ejections intermix with the occasionally infinite curvatures of space-time. Is this the order of the pleroma? Is this the reign of the one? Is it a firmament (that is, a place of fixity, anchorage, firmness, or something rendered solid—*firmare*)? Have telescopes and the theories of astrophysics truly "unified" the skies?

Perhaps this example was chosen deliberately in order to confuse the reader and to hide an imposing unity behind a cosmic "veil of ignorance." More eloquent would be the path taken by elementary particle physics, which includes: access to the microcosm, the edict of symmetries, the triumph of gauge theories, and the archetype of reductionism in action. Regularity, regulation, or wise harmony of the one resonating with itself? And yet: Within the standard model of high energy physics, the 118 fundamental elements of Mendeleev's classification have been transformed into . . . 126 degrees of freedom! Is this a reduction?

Objects are secondary. What concerns physics, one could argue, are laws. Cosmo*nomy*. This is evidently where one must search for order and unity. With the exception that symmetries are indeed shattered. That they can be shattered in several ways and that this diversity is not at all insignificant. That laws, therefore, reemerge as simple, environmental parameters: a contingency lying at the heart of formal necessity. That the increase of symmetry does not further diminish the amount of free parameters.[39] That string theory, a hypothetical convergence of knowledges that is supposed to describe the entirety of interactions and components, leads to a quasi-infinity of different, effective laws *via* the compactification of supplementary dimensions and generalized, magnetic fluxes. That universes themselves are multiplying and that a meta-strata full of diversity seems to be inscribing itself into the structure of the multiverse.[40]

In all likelihood, nothing here is accidental. This state of contemporary physics is not at all a historical singularity or a temporary diversion that would have to be corrected with the development of a more precise and astute description. Indeed, this state of affairs can be noted in different degrees in every stage of development within the natural sciences. Proliferation is endemic to this approach (does *phusis* not already contain the idea of "generation" and hence of growth?). That the desire for unification—in a very particular sense that is not *maxime unum*—has played an undeniable role in the emergence of scientific revolutions does not in any manner indicate that a physical-mathematical approach inevitably leads to a unitary real, a world unity, or a uni-verse.

The profusion of complementary theories (the electroweak model for certain interactions and the strong interaction for others;[41] quantum mechanics for the microcosm and general relativity for the macrocosm,[42] etc.) contributes to the plural, numeric, and numeral dynamic of the "more than one." This dynamic, contrary to a certain common claim, is also the prerogative—perhaps even the only possible mode of access to an authentic alterity—of the so-called "exact" sciences. A strange qualification too: If exactitude refers to conforming to the self-prescribed rules of the discipline, then it is a trivial criterion; if exactitude designates a fidelity to the truth in the correspondentists' meaning of the phrase, then it employs a very heavy epistemology that is highly debatable and evidently contradictory with the nominalist (in the sense of Goodman and Foucault), antirepresentationalist (in the sense of Rorty and Sellars), and nonmodern (in the sense of Latour and Stengers) aims or visions of scientific thought, to which we are referring here. The critiques of Thom and Vuillemin against the relevance of the concept of exact science must be extended to mathematics and theoretical physics.

Still, this dynamic of number, piling up, or extension concerns an altogether different "more than one," an irreducible and internal beat that is seemingly incommensurable with the first "more than one." It is not found in the copresence of theories whose domains of validity remain disjointed, but rather in the co-ex-istence of models that pertain to the *same* phenomena. For about a century, developing a quantum theory of gravitation has been a key scientific aim (which is not primarily related to the fantasy of a "theory of everything," but rather to the limits of inner coherence: certain places—microscopic black holes—and certain moments—the seconds following the Big Bang—seem to require this model, even if it is entirely polymorphous in the way it describes these different interactions). Strings, loops, topoi, noncommutative geometry, causal triangulation, path integrals, twistors, quantum geometrodynamics, and entropic force are many possible mediations of quantum imperatives with gravitational invariants.[43] All of these approaches essentially agree with scientific observations, or what one can approximately call observations, since facts cannot, in the words of the everyday expression, speak for themselves. (A pure "observation," conducted outside of an interpretative paradigm, is probably about as uncommon as a moral act as defined by Kantian law!) One might interpret the profusion of theories in contemporary physics as a confrontation of competing accounts that will be judged by future experiments. This is explicitly or implicitly the way that the imaginary of research practitioners works in a general and consensual manner. Nevertheless, if there is a point at which all the epistemologies converge, it is precisely that of theories' mortality, temporality, and ephemerality—not to mention their precariousness. Propositions, and this is a defining property of science, will necessarily be contradicted in the more or less distant future. Hence, at any given moment in the evolution of knowledges, there is no sense in establishing an ontic distinction between models that have not yet been disproved. As one knows, whatever sense one gives to "correction," no model can be absolutely and eternally correct. It is here, very precisely, that one must be concerned about the risk of nihilism. There are only two options: to refuse to give any credit to an approach that *cannot* agree with a certain—transcendent and transcendental—Idea of truth (which is deconstructed by Derrida's play on Freud, "[truth as] the normal prototype of the fetish");[44] or, on the other hand, to accept that in any given moment in the evolution of concepts and experiments, all of these theories are *simultaneously correct though mutually incompatible*. Following the latter approach, quantum spacetime *is* (transitively) all the models that have been evoked altogether. This position constitutes the third fissure that must be made in the myth of the physical-mathematical One.[45] There is no underlying tendency in this proposition to create paradoxes or to construct aporias. It is simply the only possible way to extract ourselves from the epistemic nihilism that scientism leads to *in fine* in all of its forms and in spite of all of its tricks. If one were to concern oneself only with absolutely correct theories in terms of a unitary meaning of truth-correctness, this would be the same as strictly disregarding all of them. As a matter of fact, the plural way becomes completely equivalent to the orthodox vision of things as soon as one has to think about the exercise of research in terms of a pragmatics or praxeology. But from an ontological viewpoint, the plural way radically diverges from this orthodox vision. The plural way establishes, through the recognition of accepted contradictions, a logic of internal displacement, a logic of the transport or push. It is repetition in both senses of the term: a resaying of the same proposition—at least in terms of corresponding to a certain "section [*coupe*] of chaos"[46]—and a preparation, a fine-tuning, and an attempt at perfection in view of the completed work (which naturally evades us, like the temptation of Tantalus). Here it is a question of stanza or *stance* and distance in a copresence.[47] It is the gap between theories that touch, caress, or scrape the same aspect of the same real—it is always a question of contact, all the way down to writing[48]—that gives rhythm to a version of the world, a version of the world that forges itself together and is fractured from within by this gap. Or more precisely: The gap brings about the beating of the ones through this fissure or crevice. This difference can only be originary: There can be no scientific thought without alternative and antagonistic descriptions. Descriptions that are accurate, we offer, and yet contrary. In other words: melodious *and yet* dissonant. Two propositions and a dissimilarity, an interval, or a deviation: The jolt of the tempo is always the same. And it is also the dynamic proper to this "more than one" that is much less heterodox than one might suppose. Indeed, the balancing—almost oscillation or undulation—between exclusive descriptions (which are repeated and repartitioned) is what builds the framework of the scientific edifice. This is not foreign to the fundamentally "differentiating" aspect of science, which offers, in the words of Lyotard, the "antimodel of a stable system."[49] The "heteromorphous nature of language games"[50]—which is internal and external—allows for the emergence of discord, which is a necessary condition for creation. What is at issue here is refining our capacity to tolerate the incommensurable whose "principle is not the expert's homology, but the inventor's paralogy."[51]

III

"More than one," as rhythm or meter, would thus be the name of our musical score. A score can, of course, be the medium for a musical text, but it can also be thought of as a border in a historical sense, as a set of sets in a mathematical sense, as a summary of statistic properties in the sense of physics, or as a division in a heraldic sense. It concerns all of these. The stakes here are considerable because the risk is immense: if "more than one," regardless of its form or metrics, slips all the way down into the heart of the hard sciences, imposes itself as a meeting place between supposedly heterogeneous philosophical traditions, and infiltrates the musical and theatrical genesis as an impulse, then can it—or must it—not be established as a principle or meta-concept?

In order to ward off this perversion (this reprise, this recuperation), there *must be* "less than one": less than one regulator, less than one universal, less than one absolute. What is at issue here is the very possibility of the immanence of sense. Of the *hic et nunc* sense, of the sense of the world as it is itself this sense, outside of any signification.[52] At issue, then, is the "subtraction from the count." Less than one does not suggest "God is dead," but rather "God cannot come back to life." Or rather: "at the limit of God."[53]

The step [*pas*] or impetus plays the role of a flexion, flux, or flow. It obeys, one could say, a complex mechanics, a fluid dynamics, or a ballistics of cataracts and turbulences. Sown with, as Michel Serres has demonstrated, circumstances, aleatory hazards, and the unexpected.[54] Fall trajectories that diverge. Shooting energy that is intrinsically local, fluctuating, and fleeting. Rehabilitated *parenklisis*, generalized *Clinamen*: *nec plus quam minimum* deviation, divided or cleaved currents and flows. A mathematics of the infinitesimal, Epicurus on Archimedes, Lucretius on Democritus. The statics of solids is not enough. Something is changing its form within material itself: movement and plasticity, *local* metrics and measures. Ever since the birth of physics.

"Less than one" does not signify "nothing." It can be an outline or a seed. A seminal touch. A gestation. It is clearly not trivial that principial or conceptual embryos—disseminated—intersect with a transitive unity's process of formation. It even results from a structural necessity, the same kind that perhaps already pervaded the paradoxical "absoluteness" of Heraclitus's exceeding. The potential hazards are the tendencies to resubsume or the attempts—both near and far—to reconstruct a global unity from the genericity of a local "more than one." The pitfall would be detecting in this "more than one" the mechanism of a new *Archē* or the foundation of another universal. "Less than one" is the praxic imperative of "more than one" as the condition for leap*s* or impetus*es*.

"Less than one" does not imply nothing, and yet it has nothing to say. Because the point is by no means to say but to diffract the already said. To make it pass through a narrow fissure that spreads out and exhibits different modes. Diffraction, as it is well known, is a self-interference. An alternation of fringes, modulated intensities, and spread-out rays. A rhythm of lights and shadows. A thought-that-means-nothing as it is announced in grammatology because it is not related to preexisting orders.[55] Besides, this thought does not create any proper order: it fashions the chaotic material from the inside. When Derrida writes in this respect at the end of a certain work that "even the [concept] of excess . . . can become suspect,"[56] not (only) excess is at issue but (also) conceptualization or conceptuality. It is the process of assembling that fixes and congeals through the "identification of the non-identical"—as Nietzsche called conceptual work—and, finally, that defuses the fissile dynamic of rhythm.

Is it necessary to evoke a "ternary essence"? Or is it not, more precisely, an effective—because it is effectuated—ternarity or ternitude? Through act more than essence or from essence to act (of the *energeia* of essence)? What is at stake in this carrying away of the one concerns the renewal of a singularity that itself must remain singular perhaps. Or constellated. The specificity of what is at stake lies in the tension between the similarity of the processes and the dissimilitude of their effectuation. The problem is not one of regularities or reiterations. It lies rather in a sort of fractality within the very mechanism of division. Must recursivity be halted? That is, by putting a stop to divisions that are deferred-different, differentiated-differentiating, so as to, with this differ*a*nce, render *the* principle undivided or even undividable? Is it the conceptual *atomos*? Questioning the very possibility of an origin or beginning (originally in writing and by extension in many other fields) was a constant preoccupation of Derrida. This questioning takes on multiple forms—and even mutates sometimes—all the way down to the deconstruction of the "center" itself as a place of irradiation and thus a point of origin.[57]

In other words, the "the counter-rule is still a rule."[58] This signifies that: "more than one" can also be the form of a particularly pregnant and totalizing "one." A new variation of the *peritropē*?[59] The Platonic refutation of Protagorean relativism is the crux of normativity in metaphysics: Relativism is, Plato insists, untenable since relativism itself constitutes a position or an absolute pro-position. Or, for what is of particular interest to us (the logic is exactly the same): "more than one" does not occur; it dissolves itself—or rather re-condenses itself—if it is itself the One. By using, in particular, some twists and turns and conundrums of modal logic, the modes that Socratic refutation has for outmaneuvering are well known.[60] Not to mention that the development of an "intolerant" relativism is committed to axiologically even though it is devoid of any contradiction.[61] This relativism is efficient, coherent, and inclusive. Today the staunchest opponents of it agree for the most part that self-refutation never took place.[62] Yet this is not, at least essentially, what must be elaborated upon here. There is another, far more efficient way to break the nihilist cycle: practice, leaving saying behind, or putting the concept itself (or what is counter to a concept) to the test—understanding, breaking through, or highlighting the dynamic of diaeresis all the way down to its own contingency.

Less than one is making sure that "more than one" does not become the new judicative authority. In the end, it is the putting into rhythm of rhythm itself. It is the displacement of displacement. The breath necessary for the step outside. Let us call this "the differents" with an "e" to remain *at the limit* of Derrida, to have to spell it—once again, more than once. "The differents" in the plural because dissemination must be incessant, everywhere, and always disseminated. "The differents," a masculine noun like *un* differend [*différend*, a difference in the sense of a dispute], because distances and repetitions—modes of the uni-finite and the uni-divided—are manifestly, indefinitely conflictual.

A. Barrau

Of Struction

I

Technology[1] supplants and supplements nature. It comes to supplant or take the place of nature wherever nature does not provide certain ends (such as a house or a bed), and it comes to supplement nature when it adds itself onto nature's ends and means. This twofold value is what Derrida inscribes into the "logic of the supplement," and one could say that this logic itself has no other source or medium than precisely this relationship between technology and nature. The supplement and its twofold concept always fall under the category of technology, artifice, or art, three words that are nearly synonymous in this regard.

Two conditions are necessary for this to be the case: to begin with, nature must present a few characteristic lacks (it is able to offer shelters, but not houses); then, it must be possible for technology to be grafted onto nature (using its materials, its forces). This is indeed the case: On the one hand, the animals of the *homo* species or varieties at least express needs that nature does not satisfy (inhabiting, warming up), and on the other hand, the technologies invented by *homo* take their operating resources (sharp stones, fire) from nature. Fire represents, perhaps, the symbolic meeting point where supplanting and supplementing occurs: It can light up during a thunderstorm, a volcanic eruption, or a spontaneous combustion of gas, and it constitutes the major "invention" of the first human beings despite the fact that it is not combustion that they invent but the conservation and "technological" production of combustion. What applies to fire also applies to electricity, semiconductors, optical fibers, and the energy that is released by atomic fission and fusion. Nature always contains and offers the prime matter for technology, whereas technology alters, transforms, and converts natural resources toward its own ends.

This very simple consideration has an important consequence: Technology does not come from outside of nature. It has a place within nature, and furthermore, if nature is defined as what achieves its own ends by itself, then technology too must be defined as one of nature's ends because it is from nature that the animal capable of—or in need of—technology is born.

Technology in turn undergoes its own development: It no longer simply responds to its own shortcomings; it generates its own expectations and tries to respond to the demands that come from itself. This is what happens as soon as the artificial selection of plants and livestock is invented. What follows from this is the construction of an order that is specific to technology, a relatively autonomous order that develops new expectations and demands from out of its own possibilities. It consists not only in the assemblage of materials and forces (what are called "simple machines": lever, mill, etc.) but also in the elaboration of logics structured by a given that is itself produced in view of a new end: Good examples include the power of vapor, oil and gas, electricity, and the atom, and later cybernetics and numerical computation (immaterial givens that at once presuppose and bring about new treatments and assemblages of matter, such as with silicon or deuterium).

What profoundly instructs this development is not "the machine," as it is all too often thought. The machine does not suddenly emerge out of nowhere. It is machined itself—that is, it is conceived, elaborated, and structured by the ends that one proposes oneself. A few anecdotes about inventions that are due to chance (the observation of vapor raising the lid of a boiling pot) cannot obscure the fact that the process of technological invention is a process specific to the unfolding of aims and investigations that are oriented by this aim. We attempt to go faster and further, to cross oceans, to produce in greater quantities, to reach the enemy from afar, and so on. At one and the same time we attempt to transport more goods, make investments for this, and insure against its risks: Financial technologies are on an equal footing with nautical technologies within a development that presupposes the existence of independent and competing entrepreneurs—that is, an entire sociopolitical and juridical technology that structures the whole space of our common way of life [*la vie commune*].

Thus, "technology" itself is not limited only to the order of "technologies" in the sense that one speaks of them today. Technology is a structuration of ends—it is a thought, a culture, or a civilization, however one wants to word it—of the indefinite construction of complexes of ends that are always more ramified, intertwined, and combined, but, above all, of ends that are characterized by the constant redevelopment of their own constructions. The transmission of sound, image, and information without a tangible medium creates new assemblages of both apparatuses and modes of life or ways of living. The possibility of acting on certain diseases or on fertility or life spans through interventions and substances that are invented for these purposes or ends creates new social, sexual, and affective conditions.

At this stage or level, ends and means never stop changing roles with one another. Technology develops a general regime of inventing ends that are themselves thought from the perspective of means (How can sterility be overcome? How can an animated image be transmitted?), and by consequence, that are thought from the perspective of means that are taken as ends (it's good to live longer, it's good that money yields more money). This is also why the technologies of the arts—that is, technologies as "arts" or the enjoyment of ends in themselves, or forms that have value on their own—can become, on the one hand, the highest standard of every relationship to ends (everything must be put into image, sound, rhythm, everything must be hypostasized into a monstration: bodies, products, and places), and on the other hand, the privileged domain for an interrogation into finality (Why art? What is it for?) that becomes suspicious of identity (What is art? What is it in the service of?).

Construction and deconstruction are closely interconnected with one another. What is constructed according to a logic of ends and means is deconstructed when it comes into contact with the outermost edge where ends reveal themselves to be endless and where means, for their part, reveal themselves to be temporary ends that generate new possibilities for construction. The automobile has given birth to the highway, which has given birth to new modes and norms of transport. It is also forcing the city to reinvent both its means of transportation (streetcars, etc.) and, over time, the very aims or ends of a "city." Digital cameras and editing processes are deconstructing and reconstructing not only the formal landscape of cinema but also the signification and the stakes of this art form (along with digital audio processing).

II

What is at stake more generally in this process is sense: Whereas we were in the habit of relating sense to an ultimate purpose or final end (whether it was one of history, wisdom, or salvation), today we are discovering that ends are proliferating at the same time as they are constantly transforming themselves into means. In this regard one could say that technology and nihilism go together: Whereas until now one used to describe ends (values, ideals, and senses) as being destitute, today ends are multiplying indefinitely at the same time as they are showing themselves more and more to be substitutable and of equal value.

Still, it is precisely here that technology conveys its lesson: Through technology, nature itself—from which technology is descended—reveals that nature is by itself devoid of an end. We knew this and we said that "the rose is without a why / it flowers simply because it flowers." But this "without why" continued to foster a more or less muted, more or less latent relationship with a hidden reign in which things were gratuitous, a hidden reign in which we thought we might be able to recognize a pure glory of

Being (as long as we no longer needed to locate a divine goodness in it anymore).

Technology teaches us to do away with this glory and hidden reign. This is troubling not only for our metaphysical, theological, and spiritual tendencies but also for our poetic ones. In a sense, this challenges all of our loftiness, sublimities, inclinations, and dispositions that are oriented toward grandeur and thus toward something other than the always mediocre measure of a life that is subjected to a necessity or need that nothing can ever explain. And if it is not explained, this need, this simple need to live, is transformed into a servitude, whereby we feel that we are slaves to technology and to its manifest corollary: capitalism, as the infinite production of values that are producible, exchangeable, and liable to grow exponentially. Value as monetary value in a way represents an inversion of nature: that which grows by itself but whose flourishing is confounded with indefinite growth and yet displays neither flowering nor fruit. "Yield" is not a random term used to speak about the profitability of an investment, including a purely financial investment (in sum, development in itself in its pure form and trade in its pure form without any reference outside of itself).

Capitalism constitutes the exhibiting of a proliferation through value—the proliferating infinity of ends and sense to which technology has introduced us. This exhibiting defines end, sense, and value precisely as the very process of an endless increase (we speak of "growth"). It is from this process that we could, as Marx did, look toward a passage through the limit and a reversal through which growth would reach a stage where its fruits would become available to all without relying on a distortion between the conditions of their production and their actual value (their pleasant taste, their value, their nontradable sense). This expectation presupposed something like a nature that would come to reclaim its rights. A *phusis* that, through technology as growth—revealing that all technology is growth—would bring about the flowering and yield of a value or sense that is free from any measure, equivalence, or possibility of subtraction or accumulation.

Yet it is not a *phusis* that is unfolding beneath our eyes. We would claim that it is the contrary of a *phusis*, and we are prepared to call this contrary "technology." Still, as I have mentioned, if technology is the unfolding of nature, one cannot see nature as the contrary of technology—or else we have to know how to consider this in terms of a reversal of nature in and of itself: But would this not renew a dialectic from out of which we would inevitably anticipate a second nature?

It is therefore necessary to think otherwise. If "technology" gives a sense to "nature," from which technology is constructed and that it destroys at the same time, this implies that speaking of nature is no longer entirely possible, nor by consequence is it possible to speak of "technology." The opposition of *phusis* and *technē*, the use of which Aristotle established, has undergone several centuries of maturation, which has complicated this opposition by contorting it in a decisive way through the introduction of what Derrida would later call the "supplement" and what Heidegger designated as "the last sending of Being." In any case, what is at stake is this: "Technology," as that which adds to "nature" and opens ends that it ignores, constructs in reality the very idea of this "nature"—its immanence, autofinality, and law of blossoming. Yet it is also nature that destroys and deconstructs this idea, and with it an entire structure of representations that have organized Western thought.

It is remarkable that the motif of destruction punctuates the dawn of modernity: first, with Baudelaire, for whom "Destruction," in his poem by the same name, concentrates all the "repugnant" and "demonic" desire that overwhelms him as he overwhelms (in "Meditation") "the vile multitude," and then, as is well known, with Mallarmé, for whom destruction was "[his] Beatrice."[2] One may also recall Rimbaud: "Is it possible to become ecstatic amid destruction, rejuvenate oneself through cruelty!"[3]

(Before the dawn of modernity, the motif of *ruin* already occupied an ambivalent place by exhibiting the melancholic charm of broken-down constructions, that is, monuments to their own ruin).

III

There has thus been something like an enlargement of construction: not so much the edification or erection of buildings, for which the temple, the palace, and the tomb formed the triple paradigm, but the montage, assemblage, and composition of forces whereby the "engineering structure" [*ouvrage d'art*] almost gives it its concept (bridge, pier, fort, hall, etc.). The engineering structure requires an engineer more than a builder, a constructor more than a founder (and incidentally, one also *constructs* roads, vessels, silos, chariots, and machines). Construction becomes dominant when edification, on the one hand, and making, on the other hand, become industrial and engineered, or in other words, when they bring into play the construction of operational, dynamic, and energy-producing schemata serving ends that are themselves invented and constructed according to defined aims (production power, speed, durability, reproducibility, etc.).

The constructive paradigm that has been spread through urbanization, means of exploration and transportation, and the mobilization of nonmanifest energies (coal, gas, oil, electricity, magnetism, digital computation, etc.)—a paradigm that has rendered ends and means more and more consubstantial—has led to a response of destruction. This does not concern ruining and demolishing so much as it concerns detaching oneself from what could be called "constructivism" (if one reappropriates a term whose invention at the beginning of the twentieth century is nevertheless not insignificant). The Heideggerian *Destruktion* of ontology, which expressly distinguishes itself from demolition (*Zerstörung*), is "destruction" in this sense (Granel and Derrida translate it as "deconstruction"). In a way it gives a philosophical counterpart to the existential and aesthetic Destructions of Baudelaire and Mallarmé. Construction as such is brought into play (as well as "instruction," as what puts knowledge into an order: one could demonstrate it through the recent use of the term "instruction" in school contexts—the expression "*Instruction publique*" [Public Education] dates back to the French Revolution and "*instruction religieuse*" [religious education] is not any older than this).

Onto what does destruction open? Perhaps onto the very movement of modern construction? What is of concern is not to "re-construct" (contrary to the incessantly repeated petition addressed to "deconstructionists": will you reconstruct already?). Nor is it to return to founding, building, constituting, or instituting gestures, even if it is to open and inaugurate, to allow for a birth of sense. What is at stake beyond construction and deconstruction is *struction* as such.[4]

Struo signifies "to amass," "to heap." It is truly not a question of order or organization that is implied by *con-* and in-struction. It is the heap, the non-assembled ensemble. Surely, it is contiguity and copresence, but without a principle of coordination.

By speaking of "nature," we used to presuppose or rather superimpose a coordination that was proper and immanent to the profusion of beings (a spontaneous or rather divine construction). With "technology," we used to presuppose a coordination that was ruled or regulated by ends that were particular to "humankind" (their needs, capacities, and expectations). By acting retroactively, if one may say so, onto "nature" from where it comes out of or emerges (we cannot decide between these two concepts . . .), "technology" muddles the two possibilities for coordination. It invites the consideration of a struction: the uncoordinated simultaneity of things or beings, the contingency of their belonging together, the dispersion of profusions of aspects, species, forces, forms, tensions, and intentions (instincts, drives, inclinations, and momentums). In this profusion, no order is valued more than the others: they all—instincts, responses, irritabilities, connectivities, equilibriums, catalyses, metabolisms—seem destined to collide or dissolve into one another or to be confused with one another.

Whereas the paradigm had been architectural, and consequently architectonic in a more metaphysical way, it then became more structural—a composition, surely, an assembling, but without constructive finality—and finally structional, meaning relative to an assembling that is labile, disordered, aggregated, or amalgamated rather than conjoined, reunited, paired with, or associated.

In fact, it is the question of a "sociation" in general that is posed alongside struction. Can there be an association, a society—if the *socius* is the one who "goes with" or "accompanies" and if, as a result, she or he brings into play an active or positive value of the "with" or *cum* around which or through which something akin to a sharing plays out? What I am calling here "struction" would be the state of the "with" deprived of the value of sharing, bringing into play only simple contiguity and its contingency. It may be, to take back the terms that Heidegger wants to distinguish in his approach to the "with" (the *mit* in the *Mitdasein* as the ontological constitution of the existent), a "with" that is uniquely categorial and not existential: the pure and simple juxtaposition that does not make sense.

IV

Perhaps struction is the lesson of technology—a construction–deconstruction of the ensemble of beings without any distinction between "nature" and "art"—insofar as it instructs us with this instruction (which, indeed, we do not comprehend, and which appears badly constructed to us). Following this instruction, sense from now on will not let itself be constructed or instructed. What is given to us only consists in the juxtaposition and simultaneity of a copresence in which the *co-* does not bear any particular value other than that of contiguity or juxtaposition within the limits according to which the universe itself is given. At the same time, these limits themselves are only given with the caveat that it is impossible to properly assign them as delimitations of a world in relation to what is beyond or behind it. On the one hand, the universe is said to be expanding at the same time as it is finite; on the other hand, it cannot even be called a "universe" but only a "multiverse." And yet, in order to think beyond the "universe," it is, of course, no longer necessary to understand the multiple worlds as one (or several) other world(s). "They are not somewhere else but modes of relating to what is 'outside-of-itself.' "[5]

The idea of the universe contains a schema of construction or architecture: a basis, a foundation, and a substruction (a word that is also found in the work of Mallarmé!) that forms the base on which uni-totality is erected and assembled. Uni-totality is posited on the basis of its own supposition and refers essentially to itself; in short, it is in itself (and "Being" is Being "in itself" within the thought that is sustained by this schema). But copresence and coappearance both turn away from the in-itself and construction: "Being" is no longer in itself, but rather contiguity, contact, tension, distortion, crossing, and assemblage. "Being," of course, shows traits of "construction" understood as mutual disposition and mutual distribution of the multiverses that belong to each other, but not as a (sup)position of a Being or a fundamental real.[6] The real does not dissolve itself at all in unreality, but opens onto the reality of its nonsupposition [*insupposition*]. This is what is signified by the dissolution of the *technē/phusis* opposition or what we call "the reign of technology."

This is what has occurred in our history. We have come to a point in which architectonics and architecture—understood as the determinations of an essential construction or essence as construction—no longer have value. They have worn themselves out by themselves.

Still, it has not only been a question of being worn out. It is not only a construction that has been destroyed by time. It is the very principle of construction that has been weakened.

The accumulation, noted above, of motifs of destruction at that time—around 1900, which is traditionally thought of as "the" turn of the century *par excellence*, the time in which in fact something was inverted and overturned, where an edifice was weakened to the point that one could say, in every possible sense, that the edifying and the edified trembled—this accumulation bears witness to a sort of saturation point and a rupture in the model of "construction." This signifies that construction bore within itself the seed of deconstruction. What first presented itself as the extension of the assemblage and montage of *tools*—continuations of bodies and simple machines—and later as the expansion of a gesture of mastery or command—the administration and governance of energies (vapor, electricity, chemical reactions) in lieu of the mere use of forces (moving water, winds, gravity)—revealed another nature: one of combination, interaction, and, later, feedback.

In reality, an entire *organicity* or a quasi-organicity has been developed. In sum, the constructive paradigm is overcoming itself; it is overconstructing itself by tending toward an organic autonomy. Overconstruction is turning into struction.

V

Or rather, according to another, slightly different perspective, it is the organic autonomy of our own behavior that has been extended very far beyond not only our bodies but even our minds by asking the latter to export and expose itself under the form of highly self-referential "machines" whose laws and schemas of organization require certain operations from our behavior in return. We learn how to use a computer, on our desk as well as in our car, in the train, on a plane, on a boat, for archaeological excavations and for recording data, and in the "creation" of sounds and images. This use not only implies a new domain of expertise but also a different space–time that incidentally is nonhomogenous and nonunitary or "universal": we are, at each moment and all at once, in the extension of certain modules that are put into operation everywhere (a digital procedure, a use of signals or icons) and also in the renewal of unprecedented possibilities, which are without a doubt very repetitive (everyone takes the same photos of the same monuments, etc.) but whose very repetition lights up a new reality. We are no longer in the process of discovering a world that has remained in part unknown; we are in a spiraling, growing pile of pieces, parts, zones, fragments, slivers, particles, elements, outlines, seeds, kernels, clusters, points, meters, knots, arborescences, projections, proliferations, and dispersions according to which we are now more than ever taken hold of, interwoven into, absorbed into, and dislodged from a prodigious mass that is unstable, moving, plastic, and metamorphic, a mass that renders the distinction between "subject" and "object" or between "man" and "nature" or "world" less and less possible for us.

In fact, we are perhaps no longer within a world or "in the world" [*au monde*]. What is disappearing or being diluted is the more advanced sense of the *cosmos* or beautiful unity that is composed according to a superior order that directs it and that it also reflects. Our "world"—or our element—is instead composed of bits and pieces that, taken all together, are proliferated from the same source (humankind, the technological animal of nature, the constructive appendage of a great all that shows itself to be rarely constructed but incredibly rich in con-de-in-structive potentialities). Still the bits and pieces or "elements"—which are never elementary enough—of this great "element"—in the sense of a milieu or an ecosystem that is an *ecotechnology*—constantly escape the grasp of every construction. Their assemblage does not refer to a first or final construction but to a kind of continuous creation where what is constantly rekindled and renewed is the very possibility of the world—or rather the multiplicity of worlds.

In this sense, struction opens less onto a past or future and more onto a present that is never really accomplished in presence. It opens onto a temporality that definitely cannot correspond to a linear diachrony. Within this temporality is something synchronic, which is not so much a cut across diachrony as it is a mode of uniting the segments of traditional time, which is the very unity of the present as it is *presenting itself*, as it is arriving, taking place, or coming about. This *coming about* is the time of struction: an event whose significance is not only that of the unexpected or inaugural—not only the significance of rupture or regeneration in the timeline—but also the significance of the passage, of ephemerality intermixed with eternity.

There is something outside of time at the heart of time: surely nothing else but what was perceived in all of our chronic thought in how time flies or gets away from us, or in the present instant's perpetual flight. Still, here, "flight" does not signify a disappearance any more than the event signifies an appearance. As with (de)(con)struction, it is necessary to uncouple (dis)(ap)pearance. "Pearance" or appearance is the appearing—but not as the manifestation of a phenomenon or as the semblance of appearance. As it is suggested by the former use of the word, "appearing" is coming into presence, presenting itself or oneself—coming near to or beside. It is always appearing with.

Within this appearing with, a displacement is revealed, a curve in the phenomenological apparatus. It concerns not so much the relationship between an aim and its fulfillment, as it does the correlation of appearing between themselves. It is not so much about a subject and a world as it is about references that send the world back into itself and to itself, about the profusion of these referrals and the way that they thus create what could be called a sense, a sense of the world that is nothing other than its appearing with: that there is a world, and all that is in the world, and not nothing.

VI

This kind of brute obviousness might seem to bring us back to a nascent, infantile, and rudimentary state. We would have nothing else to receive, project, or express but the crudest of conditions. We could not account for the world or give any kind of justice to the fact of its existence. Technology would have withdrawn any kind of final aim or end or supreme good and also rendered reason to be proliferating, exorbitant, and even delirious in its very self-sufficiency—growing like a cancer.

However, to have arrived at the state of struction does not necessarily signify having regressed or degenerated. There may be progress in the passage beyond the processes of construction, instruction, and destruction. Struction is liberation from the obsession that wants to think the real or

Being under a schema of construction and that thus exhausts itself in the pointless quest for an architect or mechanic of the world.

Struction offers a dis-order that is neither the contrary nor the destruction or ruin of order: It is situated somewhere else in what we call contingency, fortuity, dispersion, or errancy, which could equally be called surprise, invention, chance, meeting, or passage. It is nothing but the copresence or, better yet, the appearing-together of all that appears, that is, of all that is.

That which is, in effect, does not appear from out of a Being in itself. Being is itself appearing; it is appearing in an integral way. Nothing comes before or follows the "phenomenon" that is Being itself. Being itself is therefore not at all beings since it is the appearing of a being that "is" only appearing and appearing with. Thus, in addition one must say that everything appears-through together: Everything refers back to everything and thus everything shows itself through everything. Without end—and more precisely, without beginning or end.

Can we learn the logic—the ontology, the mythology, or the atheology, if one has to find a name for it—of this simple and inextricable appearing with? That is, of this *ecotechnology* that our ecologies and economies have already become, namely, states of equilibrium in our milieus and ways of managing our subsistence?

Technology presents us from all sides with dispersion, often irritation, and always the indefinite multiplication of its aims or ends that are neither ends nor means. We prolong life merely to prolong it. We manage services for these prolonged lives. We increase our biochemical and biomechanical know-how, from which we extract new possibilities for further modes of assisting other endangered lives—and we are always further away from knowing how to think about "life," not only the existence of each and every one but also the life of the ensemble of the living or of all of the living together. We are always further away from thinking nothing less than the impetus of the world through the question of "life," that is, if "life" itself—what we thus call life—is not contained within the movement of assemblages, combinations, or actions and reactions that we call "matter." Matter proves itself more and more thanks to exploratory technologies that are increasingly precise, but are themselves becoming intricately connected to their "objects."[7]

Ultimately, all that we have called "matter" and "life" as well as "nature," "god," "history," and "humankind" has fallen into the same grave. The "death of God" is indeed precisely the death of all of these substances-subjects. As with the former death, the latter deaths are very long and, in our perception(s) and even for our imagination(s), never-ending. And furthermore, they carry within themselves formerly unseen potentialities of a practical, concrete death of the living, a death of human beings and why not of the world? With each step taken by technology, not only ends, means, and deviations become indistinguishable, but harms and benefits also become intermixed, all the more so because we often do not even know what must truly be considered a harm or a benefit (for example, is the speed [*vitesse*] of transportation or transmission a "good" or a "bad" thing and according to what criteria?).

As soon as we think that we still have a few principles or rules of conduct—and, in fact, we do have some elementary ones such as bare or "vital necessities"—we cannot avoid being led toward the questions of their foundations or ultimate aims or ends. A decent life, yes, but to what end? And to which "decency"? To which level beyond mere survival? To an equality, yes, but to an equality of what if one were to go beyond the bare minimum of law? To consider each human being as an end and not exclusively as a means? Yes, but according to what? How are each an "end"? How and from where do all of the agents and levers that reduce it to the state of means enter (there are so many: economic, political, religious, and ideological ones)?

Yet we cannot presuppose that the entire assemblage and becoming of the world answers, beneath appearances that are so problematic and even aporetic, to an *intelligent design*. This idea is the typical product of a lack of thought concerning technology: It places back before nature the very *technē* that this presumed nature ends up producing.

One could also wonder whether the Western transformation, which was a technological transformation (iron, currency, alphabet, law) at the same time as it was a religious one (the end of human sacrifice, the end of theocratic empires), did not also open up the double possibility of a god that is conceived of as the one who conceives and architect of the world, and also a god who is given in distance and nonpresence. The other cosmogonies rarely if ever possess the character of a blueprint and a construction. Instead, their gods are present and active in a world in which they are, in a way, "nature" itself.

In any case, it is indeed the image of a god as an architect or clockmaker or as a constructor and technician that has emerged within and imposed itself on our culture, a Platonic demiurge combined with an all-powerfulness that took over or was put in charge of the totality of a world whose beginning and end were clearly outside of itself and in the power and glory of a "Supreme Constructor." This Constructor precipitated along with its fall a distant, personal, and living divinity of which it was the double. Thus, at the same time as it became less and less possible to understand the technological blueprint of the construction of a world (which was the question of theodicy as a justification for the work of the divine), it also became less and less possible to resort to a "salvation" and a "grace" or a "love" that ultimately would supplant and supplement an impossible legitimization.

Neither providence nor promise: One could say that it is the entire situation or situation of togetherness that is developed by technology. It is clear that any representation of an *intelligent design* is bound to fail since the "intelligence" within it only represents itself—in other words, essentially a technological intelligence or an intelligence that is purely focused on technology.[8] This intelligence can only be presupposed by its own production. Still it is condemned then to presuppose its own limits as well: Because if it is a designer that conceived and constructed (they amount to the same thing) matter and life, both of which open onto human intelligence, why does human intelligence understand nothing about why it is there once its intellect itself compels it to renounce the projections of an "end," a "second nature," "nature" itself, and a "rational" or "total person"?

At the time when a technology (pottery, architecture, clockmaking, etc.) could have been a model for the intelligent design or intent of a Prime Technician, the model implied an aim toward an end. Today the model itself—"technology" thought of as a dimension that is anthropological, cosmological, and ontological (and no longer as an order subordinated to what used to be called the "mechanic arts")—manifests itself as a proliferation or even a pulverization of "ends" that cannot possibly be imprinted onto the schema of a supposed Designer anymore.

We must dispense with "intelligent design" or intent. This cannot be disputed. One might want to argue to the contrary that a Primordial Intelligence is far more vast than ours and that its intent is precisely to make us search for, fumble around, and stumble around in the limits of the erratic proliferation of its endless goals or finalities—something like what Derrida called "destinerrance." But even if one admitted this, one would still have to face the question of an intent and design that is put to work in the wandering or errancy that we are. One could say then that the hypothesis of *intelligent design* annuls itself in another way: After having once been a hypothesis that was incapable of understanding itself, it has become a hypothesis that asks in turn for another hypothesis, a hypothesis about the sense of errancy, and even more precisely, about the sense of the errancy of sense.

To this, one must also add the following: We are not only living technicians perplexed by the development of their art or know-how. We are not only overwhelmed and disconcerted that all of the forms and aspects of sense have been brought into play and called into question. We are also ourselves already caught up in this transformation. We have been inserting ourselves into a technosphere, which is our development; what we call "technology" exceeds the entire order of tools, instruments, and machines. It does not concern what is possible through command or mastery (a means to an end), but the expansion of the brain (if one wants to call it this) within a network of "intelligence" that extrapolates a mastery that is significant by itself and for itself, a mastery that is an end and a means in itself indefinitely.

Since it is pointless to cast a veil over the errancy in struction—the veil of any preconceived "sense" that is taken from a model of "intelligence" that is supposedly "good"—then it is incumbent on us to reinvent everything beginning with "sense." Sense no longer corresponds to a schema of construction or to one of destruction and reconstruction: It must correspond to a "destinerrance," which signifies that even though we are not going toward any term or limit—as a result of providence, tragic destiny, or fabricated history—we are still not devoid of "going." We are not devoid of advancing, roaming, crossing, and also experiencing [*faire l'expérience*], a word that formerly expressed "going to the very end, to the outermost limit."

Wisdom cries out from all sides: "This must stop at once! How far will it go?" This is because, in effect, it is limitlessness that is sprouting up on all sides. It is cropping up in genetic manipulations and in financial markets, in networks and poverties, and social and technological pathologies. It cannot be a question of establishing limits for what, in itself, ignores the limit. Either this limitlessness will be self-destructive—a construction that goes up but only to fall down right at the end—or we will find a way to recognize "sense" in struction—at the place where there is neither end, nor means, nor assembly, nor disassembly, nor top, nor bottom, nor east, nor west. But merely an all together.

J.-L. Nancy

. . . And of Unistruction

> For Nietzsche the representation of the totality of the world as "chaos" is to engineer a defense against the "humanization" of being as a whole. Humanization includes both the moral explanation of the world as the result of a creator's resolve and a technical explanation pertaining to it which appeals to the actions of some grand craftsman (the demiurge).
>
> —HEIDEGGER ON NIETZSCHE[1]

I

What is revealed upon uncovering these architectural and then structural paradigms is first and foremost the great Western passion for order.[2] The worship of the assemblage, the adoration of the organized, the cult of classification. Of course, it is always possible to read and reread the tradition through any lens whatsoever and to pick up on a certain theme in it: the domination of logocentrism, the omnipotence of ontology, the forgetting of Being, the regime of the speculative, the primacy of representation, the supremacy of the ethical, and so on. It would also be just as pointless and presumptuous to search for the authentic axis or developmental objective of a remarkably polymorphous culture. Evidently, nothing can or has to legitimize or delegitimize, *a priori*, approaches or viewpoints that are significant in their consequences and (con)structions. Without making all of its ramifications, pulsions, or tensions depend on it or be engulfed by it, we run the risk of formulating a hypothesis about thought: namely, that thought—well beyond just metaphysics or philosophy—has developed for at least twenty-five centuries through an obsessive relationship to order. Or perhaps more precisely: in a recurrent phobia of dis-order. If this conjecture is correct or at least significant, the task incumbent upon us—to enter into chaos to the point of burying within it our longings for composition and our pinings for harmony or hierarchy, even it were only for an instant—is both immense and radical. It is not a question of founding (otherwise, somewhere else, or far away) but of foundering. Of dissolving the grid, of touching the amorphous. Of infiltrating into it and thinking oneself within it.

It is not of course a question of recounting all the innumerable ways in which the myth of order is the backbone of our history.[3] And well beyond the West alone, how the animistic, totemic, or analogical traditions have been innervated by this myth in extraordinarily different and sometimes even divergent ways. To consider this movement globally and generically would be to forget the point to which the concept of order—supposing that this concept is definable or even thinkable—is polysemantic itself. The order in the archaic epic poems is not the same as the one in the *Timaeus*, which is also different from that of Descartes and even more so from Bergson's "agreement between subject and object." To mistake one for the other would be senseless. But to insist that they have nothing in common or no "common measure" [*commune mesure*]—whatever the space would be in which measure would have to be defined—seems to be just as senseless. *Ordo*, *ordinis*, the file, the alignment, the row, or the line is related to the technical term *ordiri* or "to weave." It is a question of interweaving, intertwining, or interlacing according to a motif, organization, or structure. In this sense (which, strictly speaking, is not a sense because the strands or threads are crisscrossed to be precise), a certain continuity between these scattered and dispersed "orders" is delineated ("designed" or "aimed at") in spite of everything. The continuity is apparent, at least in retrospect.

What seems to us to be of utmost importance is to underscore the fundamental and even founding role that the idea (explicit, implicit, or simply latent) of a natural order plays—a natural order that is *inherent* and independent from discourse or, what amounts to the same thing, common to all discourses. A sort of generalized Kosmo-logism. Disseminated from beyond or beneath a foundation whose very legitimacy would have to have been assured or secured, an order, an absolute-altogether-order, that has transformed itself into an axiom of thought. A premise and a preamble. This order has infiltrated every crevice of the real. This is revealed in an exemplary way by the antagonistic, and nevertheless converging,[4] conceptions of order in the thought of Leibniz and Kant. "[G]od does nothing out of order. So, whatever passes for extraordinary is only so in relation to some particular order established among creatures. For, in relation to the universal order, everything conforms to it."[5] Leibniz's clarification is clear and irrevocable: there is nothing "out[side] of order." In other words, order is not simply connected to being; it *is* being. Related to his project of theodicy, the Leibnizian order integrates all the local disharmonies and discordances into itself, disharmonies that are thus reinterpreted as signs or traces of a superior reign whose subtle complexity is elusive. In sum, the stated disorder plays the role of a supplementary proof in favor of a universal order that is only partially decipherable. A magisterial loop that repeatedly or recursively closes every possible route of extraction from it.

On the contrary, one can replace or supplant divine supervision with human reason. The Kantian path searches "within ourselves," in our understanding, for the "supreme legislation of nature." The categories are functions of order (they allow, in Kantian terminology, to *categorize*). No nature is independent of transcendental ordering or not subordinated to the *a priori* conditions of knowledge. If the laws of nature could be, at least in part, apprehended independently of experience, it would be because all the phenomena are connected *via* a regulation that is already present in the principles.

Under different variations and according to diverse implementations, the Leibnizian and the Kantian "solutions" nurture the ensemble of relations-to-order. They transform this relation into the only possible or thinkable being-in-the-worlds. One cannot, or so it seems, weave together a real unless one follows certain schemata . . . They impose the existence of a meta-order to which minor orders—or local disorders—may only be subordinated. The obvious and the odd are simply emanations. Or accidents. Kant submits all phenomena to a unitary, eventually subdivided legality. Leibniz fuses and amalgamates all rules and regularities so as to derive them from a universal order. The knot is tied tightly. Turned back upon itself. All-encompassing, without an outside.

All of the superimpositions of these two symmetrical positions—with their arbitrary discrepancies and fluctuations—are possible and have indeed been explored. These superimpositions establish order not as a goal or an attribute, but as a condition of possibility. There is no thought or matter outside of order. Anti-chaos is the form of the world for *kosmos* as well as for *mundus*. Before any organizations or associations, the world is caught up in creation (or perhaps in what allows for creation).

What developed alongside the discovery of this inherent order was every conceivable mode of creation for noninherent orders, which are constituted by humans and *constructed* deliberately. Philosophy has played the role of a regulator. "Philosophy as a discipline thus sees itself as the attempt to underwrite or debunk claims to knowledge made by science, morality, art, or religion," Rorty wrote (ironically) in the introduction to his enterprise for the revocation of every *theory of knowledge*.[6] It is a question of circumscribing, delimiting, closing off, or fencing in—in a word: of ordering. But to what? Everything is possible, all is permitted, and anything can be envisioned. The diversity is clearly immense and nearly infinite. And fortunately so. Still the *order-word* remains. Like a scattered or ubiquitous density. Artificial order, which completes natural order, seems to supplant and supplement natural order's incompleteness and reinforce and consolidate it at the place where a possible breach is feared. Multiple constructions—whether difficult or effortless—to break away from disorder and to avert the emptiness that disorder seems to inevitably convey. To annihilate nihilism?

Still, if technology comes to ruin or demolish the dichotomy between *phusis* and *technē* by itself, then this partition no longer holds firm. It cannot hold firm anymore. What has broken down is the mechanism for the elaboration and legitimation of a reflection-order or a mirror-order, an image or a rough sketch, a copy or a revised version of a cosmic "grand order." The gears turn without any traction. Or more precisely: the axis of the motor is not turning anymore. A curious *dialectic of the one and order* has been installed quite possibly as a result of the intuition or anticipation of this fissure. When order no longer holds firm—or can no longer impose itself without a justification for its preeminence—the one takes its place. The one does not even need to be ordered; it produces order in itself. In itself and by itself. There is no chaos without number, no dis-order without multiplicity. Conversely, when the one is no longer, when it is multiplied or split in two, when it is scattered or reproduced, order halts this profusion. When plurality threatens, taxonomy recuperates, reduces, and recenters. It reframes (a *parergon*, once again).

According to the second logic (the appeal for order or the call to order for the purpose of averting profusion), one could mention—less as an exemplary model than as a standard or benchmark—the almost exaggerated way in which physics has indexed the diversity of phenomena or occurrences to an order that is sometimes incommensurable with what it shows or demonstrates. When the one absconds, a new reign invents and imposes itself. Chaotic systems—which fissure the paradigm of integrable systems of classical mechanics—present such a sensibility to initial conditions that the final states seem to be irreducibly *plural.* Becomings diffuse. All the way down to circumstances as simple as three celestial bodies in gravitational interaction. Probability measures, attractors, fractal dimensions . . . the theory of chaos creates new invariants. It categorizes. It orders so as to overcome the exuberant and the numeric. It is almost as if the withdrawn or scattered one is resynthesized, amalgamated, and conglomerated by the stated–decreed–invented order. It is not a matter of criticizing, refuting, or discrediting this approach. Its scientific legitimacy is indisputable (physics' study of chaos clearly constitutes one of the major advances of the twentieth century). It is rather a question of underscoring the persistent logic of order according to which this approach is developed all the way down to the description–inscription of those phenomena that are most notably opposed to being subsumed by the One.

Beyond experimental results, phenomena, objects, and events, laws themselves seem to be inscribed into a plural real. Their diversity is, once again, evidently *irreducible*, at least for low energy states. Attempts at unification concern a "high energy" state—associated, for example, with the first moments that followed the Big Bang. Nuclear and electromagnetic forces can be considered in effect as "unified" but fortunately that does not concern normal temperatures (if that were the case, our world would be impoverished and uniform, dull and colorless). Even worse: What is in question today is the search for a fifth (a quintessence, as one says)—and hence new—force that could explain the acceleration of cosmological expansion.[7] It seems that the unifying reduction has not quite yet been achieved . . . on the contrary! Without a doubt, a new system, a new language, a reconceptualization, or a reassemblage will permit the precipitation of these laws into a new condensate. It is of little importance whether it is called string theory, M-theory, or something that still remains to be thought. It will function as an operator of order that, in the absence of a genesis of unity, will proceed by categorizing diversity. This thought-machine is like a *weaving* loom.

Conversely, following the first logic (the appeal to the one so as to disguise or hide the lack of order), one could mention the question of the plurality of the arts, which is put forward in an exemplary and recurring way in aesthetics. It is less a cause for concern than it is a source of anxiety, less a problem of number than a turmoil that is unfailingly related to disorder. Uncategorizable, the arts belong to one another, superimpose themselves onto one another, and hybridize with one another. In the face of this constantly ruined order that is imploded by way of practice like a grafted scion, the traditional response of philosophy consists in—as Jean-Luc Nancy has shown—evoking or invoking the "essence of art."[8] Not of arts, but of *art*. In the singular. Sometimes even with a capital letter. Indeed if one were to look beyond appearances, to free oneself from the artifices of sense, to reach or touch the origin, to perceive the fundamental drive, there would only be "*one* art." When one peers into it deeply, when one excavates the ultimate signification beyond the uncategorizable ways of expressing it, the uniqueness of art *must* be revealed. When order fails, the one is erected.

The efficacy of this perfectly well-oiled dialectic confines one to limbo. The swaying back and forth is, by its very construction, guaranteed to function flawlessly. It self-compensates, self-adjusts, and self-regulates. It is an undulation or a swinging that is immune to its own movement. When the one withdraws itself, order invents-imposes itself; when disorder reigns, it is because the one is already at work, eventually hidden, crouching down, subjacent, and barely touching. The real reorganizes itself under the twofold yoke of the one and order. They are two sides of the same world, which are infinitely near to one another and yet arbitrarily distant on two opposite sides of this endless plane.

A world in which, finally, everything is *one* and everything is *order*. The one is order for itself; order cannot be anything but either the trace or sign of a one that remains to be discovered, or else the unity of ordering as such.

It is now a question of extracting oneself, precisely from this systematic deafening, from this logic that confines one to limbo. This dual imperative, which is apparently indisputable, must be contradicted or contested. Yet to deny the order of order does not come without risk.

The ruse of the tradition consists in never having discovered that this connection between the one and order, this referral without end and without other, is a construction (and that it is susceptible, therefore, to being deconstructed). Rather it discovered in this connection a cardinal rule, an inherent that is congenital to beings or conceivables. The "way of" became mixed up with "being-in." And *putting-into-operation* became ontology.

Thus, the question of the "more than one" is not heterogeneous or independent from the question of "struction." Both confront these two fundamental notions that haunt a very long and very static history of thought. These two pillars ensure and reassure. Sometimes they even decree. Less in the manner of columns or pilasters that support a revelation or Truth and more like a kind of twofold, supporting structure that forms a backbone that runs throughout the possible or conceivable becomings-in-the-world. The one of order and the order of the one are carefully enjoined like the strip and mirrors of an all-encompassing praxinoscope. To move past or to pulverize the paradigms of the one and order (which can therefore only be done in a concomitant and coordinated manner) is not an easy task. It is even less of a rhetorical exercise or a theoretical artifact. It is not even a displacement or a reversal. It does not consist in working backward or against a few of the dominant theses and hypotheses, but rather in immersing oneself, without breaching a passage, in a chaotic magma that is no longer orderable or reducible. Neither in law nor in practice.

II

Two authors have approached total disorder with a singular persistence, skirting past its flux of infinite speeds or its colorless and silent inertness, perhaps like no one before them had ever dared.[9] Deleuze and Guattari (or the conceptual machine deleuzeguattari), in the final pages of their last book together,[10] explored with a keen perspicuity the modes and strata of the possible and thinkable being-in-the-chaos(es) as that which forms—in their sections—the brain.

"In short, chaos has three daughters, depending on the plane that cuts through it: these are the *Chaoids*—art, science, and philosophy—as forms of thought or creation. We call *Chaoids* the realities produced on the planes that cut through the chaos in different ways."[11] The role of chaos is primordial (and not foundational because, as Zourabichvili demonstrated: it is *virtual*; it is thought more than given; it takes shape and vanishes; it is entirely unable to found anything).[12] Its definition is firmly linked to the definition of the plane of immanence or the plane of Nature and univocity, "which knows only longitudes and latitudes, speeds and haecceities."[13] Chaos acts like a cut or sieve. "In fact, chaos is characterized less by the absence of determinations than by the infinite speed with which they take shape and vanish. . . . Chaos is not an inert or stationary state, nor is it a chance or mixture. Chaos makes chaotic and undoes every consistency in the infinite."[14] The acknowledgment of this dynamic inside of chaos and also of its role as a source of latencies—and, if one can speak in these terms, as a reservoir of the infinite—is essential to the concept of the chaoid. It is quite remarkable and highly innovative that modes of thought are strictly defined here in relation to chaos. What is more, they are defined by the manner in which they slip into chaos. Chaos is that which "propels" thought and no longer serves as an impediment or hindrance. It becomes active or actant—almost an agent. It breathes life into [*insuffler*] the impetus for pushing toward the inaccessible. Causal succession and the level of ontic hierarchizing find themselves entirely devastated. The step [*pas*] is immense—the perspective is inverted. The proposition is amazing.

Still a fear, almost an anxiety, appears as a counterpoint to the establishment of a driving and generating chaotic primacy or primality.

Indisputably, this prompt is—and this is startling—first and foremost to penetrate into the chaos, to throw oneself into it without restraint, to drink or to become drunk on its flow of instabilities and turbulences:

"[A]rt, science, and philosophy require more: they cast planes over the chaos. These three disciplines are not like religions that invoke dynasties of gods, or the epiphany of a single god, in order to paint a firmament on the umbrella, like the figures of an *Urdoxa* from which opinions stem. Philosophy, science, and art want us to tear open the firmament and plunge into the chaos."[15]

The revolution begins. But is aborted. The following sentences correct, deviate, and decenter: "We defeat it only at this price. And thrice victorious I have crossed the Acheron."[16] (Why *three* times? Why three instead of two as in Nerval: must one therefore go back to the underworld or come from it *in fine*?). Surely there is a relationship to chaos. There is even being-through-chaos. But there is never being-for-chaos. This relationship is full of conflict and anxiety. Sometimes it is even warlike and bellicose: "We require just a little order to protect us from chaos."[17] Deleuze and Guattari invoke and call for "protective rules . . . preventing our 'fantasy' from . . . crossing the universe"[18] as they evoke an "objective antichaos."[19] (And it is from Kant and not from Leibniz that they borrow the following example: "If cinnabar were sometimes red, sometimes black, sometimes light, sometimes heavy . . . , my empirical imagination would never find opportunity when representing red color to bring to mind heavy cinnabar").[20] What is counter to chaos is a "struggle." It is a matter of its "defeat," they write. To avoid by all means "falling back" into chaos, which would be like a . . . "fall"—a word fraught with meaning! The risk is considerable since one wants to avoid being "dislocated" or "submerged." One "struggles" against chaos. One "responds" to it. It is an "abyss" that the thought-brain "confronts."

The two pitfalls for Deleuze and Guattari are chaos and opinion. The statuses of chaos and opinion are not symmetrical, however. One must borrow from the former its destructive powers and annihilating abilities and then use them against the latter in an attempt "to defeat it with tried and tested arms."[21] The struggle against opinion is more profound; it is even "more important" because the "misfortune of people comes from opinion."[22] One must "hasten the destructions" so as to make clichés ineffective.[23]

A chaotizing and chaotized ambiguity. It is clear that chaos must be defied. Still chaos is also what allows for emergences, the material from which one makes variations, varieties, or variables, depending on whether one thinks in terms of concepts, sensations, or knowledges. It is the matter that is chopped up or carved out through planes of immanence, composition, or reference.

This relationship remains fundamentally antagonistic. The issue is to bring the chaotic adversary to the ground. As Zourabichvili writes, "there cannot be an experience of chaos, since it would be nothing but the collapse of . . . thought."[24] The chaosmos of Joyce and Deleuze is not the chaosmosis of Guattari. As Guattari often mentions in his journal entries, which often carry a tone of astonishment or suspense or even almost idleness at times, what is essential for him is in effect the reopening of disorder, the whatever, "the mess," the "active forgetting"[25]: what he calls the "schizo-analytic project." Something opens up "when it doesn't work, when it spins off course, and the sentences are fucked up, and the words disintegrate, and the spelling is total mayhem."[26] Still Deleuze veers away from this and is inclined toward closing the text onto itself or toward the complete work in other words.[27] Guattari considers this to be a form of "overcoding." The global or overall coherence of their approach is evident and undeniable; both of them wish to "put together, on a page, what flows and flies off in all directions." Deleuze and Guattari are like the deterritorializing and reterritorializing aspects of the refrain. The deleuzo-guattarian project is striking: from its vitalist dynamic all the way down to its becoming minor, it *constructs* a rhizome without a precedent. It constructs because it is indeed a question of construction. One must proceed by way of "concept construction," they conclude. As Véronique Bergen writes, "In ideal genesis, the Deleuzian image of thought as a throw of the dice is inscribed into the terms of a *constructivism*, a constructivism along which thought, through its making consistent, has to wrest itself from chaos."[28]

This paradigm is not (yet) that of struction. And this is precisely the reason why, for Deleuze and Guattari, chaos (still) must present itself in the form of a rival or enemy, even if chaos is the object and subject of numerous alliances. Deleuze and Guattari probably pushed back these borders (the frame within the frame) as far as possible. They brushed up against a line of rupture (which was precisely not a line of flight); they traveled along this line like tightrope walkers and tripped over its "molecules," occasionally flowing toward Khaos and then regaining their balance through Kosmos.

Still they did not cross over the line of rupture: Structional—rather than structivist—thought has no place there. Their chaoids can only be chimerical or hybridized. Disorder was certainly not ruled out; its role is essential. It acts, impacts, and affects. Still it must be linked, organized on the other end, or connected to a ground rod. It is significant, but it is not significant *by itself.* Like all Deleuzian concepts (or what is structurally related to them), disorder is "syntagmatic," "connective," "linking," and "consistent"[29]—all things that non-order or structional post-order cannot be. Crisscrossed by intersecting planes, the deleuzo-guattarian chaos operates and shows itself to be infinitely helpful in this sense in accordance with the establishment of a "logic of the AND" that would "overthrow ontology, do away with foundations."[30] It approaches the *limit* or the *exiting* of its own system.[31] It is both the final point where constructivism self-dissolves (or exhausts itself),[32] and a form of invitation to move beyond the schemata of construction—that is, to undermine the architectonic or structural logic. Still, here again, here too, it is an *order-word.* Rather than planning a revolution (which is, actually, impossible) or spreading the conditions of rupture-scission to come, Deleuze and Guattari surveyed and then invented the end of a world. They *touched* chaos. They allowed us to envision the leap by eliminating the bridge: "There is no bridge here—only the leap."[33] They opened a window of possibility so that we would not suffocate.[34] Yet without becoming one, or one body [*faire corps*] (even without organs), with disorder.

III

To truly enter into disorder, to explore its crevices, to climb out onto its branches, to get drenched by its tumultuous downpour, perhaps one would need to return to the technological aporia that pushes one—that compels one—to no longer fear the disorderly and the irreversible, and finally, that orders or enjoins one to leave entropic phobia behind.

Plurality, though not completely aporetic, is not free from contradiction when it becomes normative or prescriptive. There *must be* "more than one." Its inner rhythm is the creative pulsion and searching pulsation. However, as we have underscored, there must be "more than one" in a sense of praxis, almost an aconceptual sense. The idea itself is almost discordant: The "more than one" as a concept or regulative principle would be erected immediately as the most dangerous and insidious "one," the one that ignores itself as such and closes itself back onto its own opening. All-encompassing and all-embracing.

It is not the same for the—different but related, external but analogous—question of struction. Thinking within the regime of dis-order is clearly not without risk. Yet the risk is of another nature. Contradiction no longer threatens us. It could not be a performative contradiction in any case! Stating [*énoncer*] disorder, fortunately, does not amount to reducing it or putting it into an assemblage. Chaos does not lead to its own negation or resolution. The leap is not followed by any wake or furrow. There is no splash after the dive. There is only an always-already present current. Removing oneself from the stronghold of order, from the widespread ordering, requires a specific clinic. *A pharmakon*?

The role that technology plays in the collapse of its own paradigm, namely, that of construction—that of exercising a mechanical "know-how" that is almost outside of nature—must be examined against its own intricate entanglement with/opposition to the thought of art.

If technology is conceived of here as that which is in opposition to Nature—all while being internal and perhaps inherent to it—then it has also been structured by its opposition to art. "Art and technology," writes Nancy, "are so distinct for us that the title 'art and technology,' which has been the theme of more than one essay and more than one exhibition, is necessarily understood as the assertion of a problem and not as a tautology."[35] Still, in pursuing this further, the thought of art and the profusion of technology "are united in opposition, by their common opposition to what we still call 'nature.' "[36] What is discussed here then is a triptych: art / nature / technology. Three terms, three couples: art and nature; art and technology; nature and technology. Six relationships if directionality is included, if the preposition is noncommutative, if asymmetry is integrated into these relationships—in other words, if there is *a sense*. Here the question is not to demonstrate once again why "[i]t might be that art, the arts, is nothing other than the second-degree exposition of technology itself, or perhaps the technology *of the foundation or ground* itself. How to produce the ground that does not produce itself: that would be the question of art, and that would be its plurality of origin."[37] By contrast, it is important to underscore that all of these couples or twos have a relationship to excess and the plural, to moving beyond the other boundary stone, to pushing it out further *ad absurdum* and hence to its dislocation. These couples or twos always operate in the perspective of the "third" term. Their interaction is a "force of three bodies." They do not constitute poles or nerve centers, but influxes, influences, or whatever produces movements. They are less the angles of a triangle than its sides. Indeed, they mark the limits of the limits. They clash into each other and implode the very building that would have underlain their connections, and this has been the case from the beginning. "Technicity *itself* is also the 'out-of-workness' [*désœuvrement*] of the work, what puts it outside itself, touching the infinite. Their technological out-of-workness incessantly *forces* the fine arts, dislodges them endlessly from aestheticizing repose. This is also why art is coming *to its end*."[38] This very same schema unfolds in the other relationships. Technology acts as a positive "feedback." This signifies that technology does not regulate but rather destabilizes. It is a principle of divergence. It pushes the system to a point of saturation. It does not invite; it imposes and implicates. Its moving beyond and out-of-control movement is relentless, a movement that is part of the technological process as the mode of the system's dissolution into its own excess. Already, a problem overcomes the apparatus down to the motif of its enactment: an excess of zeal, an overzealousness.

Still, this story does not end here. Indeed it never ends, even when it finishes or is finished: It is the unfinished-unfinishable by way of tendency, by way of tension; it is the out-growth. As Heidegger has shown, technology is also related to *unveiling*.[39] It is the very "foundation" of technology; it exceeds its sense of handicraft immediately (imperatively, one could even say). It is poietic, and perhaps even far more than this. Up until Plato, *technē* was associated with *epistēmē*. It was a question of knowing, of being knowledgeable, of finding oneself knowledgeable *about* something. Yet in a certain sense at least, and in "more than one" sense in fact, the movement is amplified and intensified with the arrival of modern technology. Heidegger's *enframing* refers to the provocative call—or provocative seizure—of technology in the era of machines, motors, and automatisms. What gets unveiled must be considered as the foundation or base. Here the key cannot be found in technology's output or products, but in its essence ("the true by way of the correct").[40] And this essence is primordial because it becomes, as Heidegger considered, a source of "unrest." A source of restlessness as well as something that makes one restless. This essence is ambiguous and polyvalent. Against all expectations, the "essential unfolding of technology" thus contains within itself "the possible rise of the saving power."[41]

Considering that from the initial pages of Heidegger's *Introduction to Metaphysics* onward "every essential form of spirit is open to ambiguity,"[42] it is not surprising that Heidegger encounters the aporia of technology and discovers or finds in it the possibility of an overcoming (or a return)—even though he notes or underscores the limits of order. Philosophizing, he writes, "is questioning about the extra-ordinary."[43] More deeply still, it is questioning itself in all of its forms and even anamorphoses, and it must be placed or displaced outside of order. This is what Heidegger calls the "recoil[ing] upon itself" within the act of questioning.[44] In other words, "Philosophizing . . . is extra-ordinary [*außer-ordentlich*] questioning about the extra-ordinary [*das Außer-ordentliche*]."[45] This text must be taken literally. First, if there is an "outside of order" or even worse an "extra-ordinary," then there is an order or even worse an ordinary. In addition to this, this "outside of" is "extra-ordinary." Order is therefore what is ordinary by default. Order is here, now, developed, present, co-present, and probably omni-present in all of this thought, which is neither philosophical nor poetic. Heidegger never ceases to underscore that this latter thought does not reach the level of "autonomy" and "creativity" that philosophical or poetic thought does. In this sense, the increasing awareness of the necessity of extracting oneself from the constructional paradigm agrees in part with the analysis (and hope) of Heidegger. Still, it does not agree with it entirely. Heidegger sought to place intelligence (human, very exclusively human intelligence . . .) into a dynamic that departs from order, but he remained dependent on the existence of an inherent natural order. It is according to this order, and in relationship to it, that motion, thought's sense of flux, is defined. Without expressing it explicitly, this idea, this inexpressible belief (or rather: this belief that is not necessary to express) is able to be seen in all of Heidegger's work:

"To be sure—whether the question 'Why are there beings at all instead of nothing?' is posed or not makes no difference whatsoever to beings themselves. The planets move in their orbits without this question. The vigor of life flows through plant and animal without this question."[46] This is very clear: the course of the planets is not a construct; it is therefore not—here once again—susceptible to being deconstructed. Hölderlin could think outside of order:

> But where there is danger
> The thing that rescues also grows.[47]

And yet order still equally determines what the speech of the poet falls on or lands on. Clearly, speech deports [*déporter*]. Heidegger extracted himself from the dialectic that keeps one in limbo between the one and order. He was able to think dis-order. Toward disorder with his back toward order. Yet he did so without breaking out of the cosmic (and indeed no longer cosmological) frame in which this exercise of elevation or transformation has to be put into practice. According to this perspective, it is of little importance that Being detaches itself from its correlates and ends up being reduced to beings, thereby confining thought to the position of a mere point of view. Nothing in this evolution—or decline—evokes the register of thinkable disorder. Surely, one must indeed, in the work of Heidegger, extract oneself from order so as to think-create. Yet by turning one's back on it, one also leans against it. It becomes a base or pedestal to stand on even if it merely provides a jumping-off point or impulse for a movement of flight. It is an overdetermination of that which one wanted to ignore. It is only a position or posture. An oblique look, a subversion perhaps. Heidegger's object-world is still supported by its own order. An untouchable, irrevocable, and invincible order.

IV

Thus, from the viewpoint of struction, neither the Deleuzian approach, which could be called touched-chaos, nor the Heideggerian gesture, which is a displacement outside of order, jam up the Leibnizian–Kantian dialectic. This meta-order was never questioned because, in the approaches of Heidegger and Deleuze, which play dangerously with its borders, it was not able to be questioned. Indeed this does not concern a reproach of Heidegger and Deleuze, which would be as senseless as it would be inadequate. Their approaches are simply *infinitely* different from the approach of struction: They confront disorder; they do not develop or unfold within it. Their aim was different. The delaying-unframing-shifting that is in question here is not that of Deleuze or Heidegger.

It is a return to the origin (to a certain origin—which is also a bringing to completion).

The *Timaeus*, traversed by both an extreme traditionalism and a stunning originality, constitutes the most intensely paradoxical ode to order. Steeped in myths (and here Plato plays with the pivotal or transitional temporality of the work so as to establish in it what one could almost call a regime of ambiguity), this work still manages to establish new rules that pose "for the first time in history," according to Luc Brisson, the problem of scientific knowledge by founding this knowledge on an "axiomatics" and thus naturally also on a mathematics. This strange position that *Timaeus* holds—a text that is marked by mystery and almost by magic at the same time as it claims to have a rigorous, causal logic and a certain verisimilitude (sometimes an *eikos muthos*, yet more frequently an *eikos logos*)—is not foreign to the text's very particular status, which is at once singular *and* universal. A text that is at the margins of the great cosmologies,[48] and yet condenses or concentrates a large portion of their common threads.

The *Timaeus* interests us here on at least two levels: the specificity of its formal and literal approach toward a perfectly ordered universe; and the manner in which (the) *khōra* has been addressed (reshaped or liberated?) by Derrida (in a sense that one could describe as being *absolutely* removed from the prefiguration of Cartesian space that Heidegger saw in *khōra*). The Derridean movement itself has two levels: what it says and what it does, what it gives and what it asks for. As analysis and as synthesis. What is revealed by these intricately entangled senses and *contresens*, or erroneous interpretations, is something like a common origin of both order and its inexorable overcoming/displacement.

It seems that everything is said in these few lines. Socrates congratulates Timaeus for his admirable prelude or opening speech and encourages him to expound upon this theme without interruption. We should note: This weaving that will put things into order is not to be inter-rupted. If it were, would it risk being unraveled? The aim of this long monologue is thus to understand the reason (in the singular) for which the one (in the singular) who made our universe (in the singular) fashioned it: "For since he wanted all things to be good and, to the best of his power, nothing to be shoddy, the god thus took over all that was visible, and, since it did not keep its peace but moved unmusically and without order, he brought it into order from disorder, since he regarded the former to be in all ways better than the latter."[49] Timaeus's explanation is simple and clear. The work of the god is the "best possible [*meilleure possible*]"; it results from a "forethought" or "decision that has already been reflected upon [*décision réfléchie*]." Order is not a human projection or a specific interpretation. Order is an integral part of the very idea of the craftsman-demiurge. Whereas the transmutation of some elements into others, the location of the parts of the human soul, and bodily diseases are described in full detail, the issue of the primacy-decree of order is all but concluded in one sentence. The order of the world is the being of the world.

Alongside this brief, but essential *putting-into-order*, a reflection on the one and the unique is developed. The *Timaeus* is devised according to this double helix: the growing helix of harmony and the collapsed circle of unity. In the very first sentence of his monologue, Timaeus evokes "our universe [*notre univers*]."[50] It is clearly not neutral to refer to *our* universe rather than to *the* universe.[51] A door opens onto the possibility of a plurality (which is not self-evident: On the contrary, Hesiod uses the word "universe" three times in his *Theogony* and implies that it is nothing but unique). Likewise, it is "our world [*notre monde*]"[52] that contains all of the visible creatures, as does living-in-itself, which contains all of the intelligible living things. Later in the text, Timaeus introduces the discourse of the demiurge as the discourse of "the begetter of this universe."[53] *This* universe and not another one whose eventuality is clearly implied. Astonishingly, in the quick—and highly ironic—recollection of divine genealogy that precedes this allusion, Plato places Gaia and Ouranos at the origin of the ancestry of the gods. Khaos has simply . . . disappeared! No more gaping void, breach, crevice, gash, or primitive fault. This element—which was until then primordial, absolutely unorganized, and prior to Earth and Sky—is not required anymore. Still what is of utmost importance here is to note that if Timaeus opened up the possibility of a multitude, he did not openly affirm it or give validity to it. This could thus not be due to carelessness or approximation: It is after having considered multiplicity that the singular is imposed. Here again, the clarification is categorical: "So have we spoken correctly in naming the heaven 'one,' or was it more correct to say that it's many and indefinite in number? One, if indeed it's been crafted in accordance with its model."[54] However, "its model" is the very model that permitted the establishment of the superiority—and consequently its efficacy, as the demiurge is good—of order over disorder. "[T]he maker did not make two or indefinitely many cosmoses; but rather this heaven here that's come to be, both is and will continue to be one—alone of its kind."[55] The mechanism is in place. The gear is ready to be turned. All of the elements of the one-order-system are already here. Watching out for any possible escape. This perfectly self-regulating and self-maintaining dynamic—one that initiated a posterity that has not yet been exhausted—installs the dialectical framework. The one and order are not the same thing. They are antinomies: if the One ruled, what would be left to put into order? And still the one and order obey the same fundamental schema. Or perhaps rather: They are this schema in its twofold texture. They are structurally identical. They develop alongside one another. They intertwine until they become indistinguishable from the weft of the real. Timaeus completes, brings to completion, brings to a close, or encloses these two openings—disorder and plurality——almost in the same way. The weft does not fray on either side. Plato went very far. He surveyed the shifting soil of dangerous lands. He took risks: he considered non-order and the multiple. He almost posed the question, but only to put it to rest or bring closure to it: The only possible "likely story" [*mythe vraisembable*] is that of a world that is unique and assembled according to a static and unchanging order. His aim was indeed also—and especially—chiefly political, and clearly his argumentative structure was just as suitable for defining stable norms that could guide the city or describe the origin of the universe (or of humankind in the *Critias*). This consistency-complicity-confluence is perfect. The rough spots to come are patched over even before they appear. This knot has continued to be woven well beyond its original scope. It is not merely the best possible route; it is the implosion of the world into a single path. A closure.

The archetype (or typology of *archē*) of this one-order is an intelligible form. Intelligible forms are pure, unmixed, not sensible, existing absolutely in themselves, and not subjected to becoming. In order to establish these, Plato appeals to three proofs, each of which has a very clear architecture: a metaphysical proof (that of the *Parmenides* and the *Phaedo*); an epistemological proof (that of the *Republic* and the *Timaeus*); and an etiological proof (that of the *Phaedo*).[56] They are conjoined according to a very precise pattern: logical homogeneity, putting into order, and a reduction into unity. Here too the same scenario is at work: turning a principle into a law (here that of noncontradiction), solidifying and stabilizing those elements that are aimed at (here, sensible forms by way of their participation in intelligible forms), and finally, unifying the real that is thus subsumed (here, the properties that are displayed in the *Phaedo* from 95e to 102b). The enclosing wall is in place and meticulously fortified. It is practically inviolable. It has no exterior. This matrix or womb has to be thought of according to its rules and lines. It defines the space of the possible.

In this mirrorlike conjoining of the one and order, the weft of the *Timaeus* portends a long tradition. Myth, ritual, and logic were certainly not solely introduced by this text and clearly have not declined in our time. And yet, beyond the establishment of this "structure," which is all-encompassing and infiltrated all at once, the narrative [*récit*] (since it is not truly a dialogue, in fact) is also very significant for what we are concerned with here through the introduction of (the) *khōra*. Is it an indeterminate meta-material that is open to receiving every creation?[57] Through (the) *khōra*, it seems to us that Plato spreads an element that is surely essential to the *construction* of the cosmos, yet an element that is also destabilizing for—or at least disruptive to—the grand order that is being announced. Just as it makes possible a "logic different from that of the *logos*,"[58] would (the) *khōra* not also be designed to open onto *a world different from that of kosmos*? In other words: Does the *Timaeus* not contain in the end, and against all expectations, the germ or seed that will dissolve its own rhetoric? Not as an aporia or internal contradiction in fact, but as a work that contains, as a mode of its own edification, the principle of its collapse. An infection of its principle. Derrida took up this concept (which is not truly a concept, since it frees itself from what Deleuze calls the "state of survey") of the *khōra*[59] to turn himself toward what it might (re) present in terms of a "third genus" (*triton genos*, as Plato writes).[60] The definite article (the "the" of "the *khōra*" to which the tradition refers) is revoked by Derrida because it implies "the existence of a thing," a being "to which, via a common name [*nom commun*, or 'common noun'], it would be easy to refer."[61] In any case, it is because of this that this work of Derrida forms, along with "Passions" and "On the Name [*Sauf le nom*]," an essay or work of fiction on the name or noun [*nom*] itself (and by consequence on the nickname, the unnameable, the aptly named, the pseudonym, the synonym, and, of course, the nominalism which is never *named* as such by Derrida but probably offers a very helpful lens through which to read his overall work). Some thing (or some non-thing, whichever one prefers) about (the) *khōra* seems to defy the dichotomy between *logos* and *muthos*. Reasoning is in another place, in another type of discourse, as if it were decoded "beyond categories, and above all beyond categorial oppositions, which in the first place allow it to be approached or said," writes Derrida.[62] But *saying* indicates *to say* well beyond saying itself, since it is a principle as much as it is a substratum. The masculine noun being the [*le*] discourse or saying, and the feminine noun being the [*la*] *khōra*? It seems that the *triton genos* is also to be understood according to its sex. A strange curiosity that one is also tempted to compare to the highly ambiguous status of Khaos, whose descendants are begotten without any coupling . . . A third genus or gender [*genre*] outside of gender [*genre*], which muddles the very possibility of naming, but founds the need to classify and order (to put into order *and* give an order). It is a surprising and intentional ambivalence. The thought of "*khōra* would trouble the very order of polarity, of polarity in general, whether dialectical or not."[63] Still the order of polarity is order itself, *in* itself. There is no order without poles to lean on or hold on to. A monopole, a dipole, a multipole: Regardless of which one, the (disciplinary or force) field lines (even when broken) always converge toward the pole or poles. "Thinking and translating here traverse the same experience. If it must be attempted, such an experience or experiment [*expérience*] is not only but of concern for a word or an atom of meaning but also for a whole tropological texture,"[64] and, one could probably add, for a whole allegorical textuality. To say it in another way (inasmuch as one can still say), Timaeus intertwines and unhinges the interpretive grids that he himself creates. The name of (the) *khōra* "is not an exact word, not a *mot juste*"[65]—the process it puts into play is not accidental but "inevitable." It "anachronizes being"[66] and anatopizes it just as much, we presume. It is the ensemble of the weft or the subjacent, skeletal structure that stops functioning as a material for fashioning or a matter for founding. Once more, more than once, it is a question of the essence's sense that is raised and in the end nullified or destroyed. It is lost without the profusion of its consequences. It is off the beaten path. Is it anaporetic? The answer is: there is no answer. Not because of a lack of recognition or completeness. Not in a transitory way or by default. Quite to the contrary: it is by scrutinizing the skin of (the) *khōra*, by analyzing it almost to the point of dissecting it, by interfacing its squamous residues with the world that it surrounds and that surrounds it, and by putting on or wearing (the) *khōra* itself that this consequential and creative negation takes shape or is delineated.

"*Khōra* is not, is above all not, is anything but a support or a subject which would *give* place by receiving or by conceiving,"[67] which signifies: It is not a matter of (or a space for) construction. It is outside of the paradigm of construction (in the precise sense that *paradeigma* has in the *Timaeus*). Perhaps to the point of destabilizing the paradigm? *Khōra* is indispensable for the "likely story" [*mythe vraisemblable*], yet it also evades the project that undergirds it. Like Khaos, it does not designate any type of known being. *Khōra* is exterior to. As Derrida writes: " 'something' is not a thing and *escapes from this order of multiplicities*" (italics mine).[68] Here indeed one reaches a breaking point and touches a point of rupture. The dialectic between the one and order no longer functions: The regime of multiplicity begins to be delineated, and then order cannot subsume this multiplicity back into itself, nor can the assemblage reframe it. It is an other of *logos* that is not a *muthos*. It is a glitch in the mechanism, a preface or footnote that would also be the body of the text and its conclusion, a quality that would not solely be measured "by the nonmythic character of its terms,"[69] and thus in the end would take up its *value of disorder*. *Khōra* is necessary to *Kosmos*, but it is also potentially fatal to it. The entire issue is all right here.

As Derrida warns: "Let us not be too hasty about bringing this chasm named *khōra* close to that chaos which also opens the yawning gulf of the abyss."[70] Indeed Khōra is not Khaos. Still Khōra is neither foreign nor unrelated to Khaos. If the encyclopedic ending of the *Timaeus* "must mark the term, the *telos*, of a *logos*, on the subject of everything that is,"[71] then one could formulate the hypothesis that this ending must also determine the *muthos* of a *khaos*. The point is not at all to presuppose the existence of secret or esoteric connections buried in the text, but to simply and meticulously put into practice this "*mise en abyme* of the discourse on *khōra*,"[72] to touch this "feeling of dizziness, on what edges, up against the inside face of what wall: chaos, chasm, *khōra*."[73] Derrida digs, bores, and excavates, but never exhumes. "The forces that are thus inhibited continue to maintain a certain disorder, some potential incoherence, and some heterogeneity in the organization of the theses"[74]: This disorder is not an accident or an obstacle; it is the mode of surpassing exegesis. It is what in this text, at the very heart of the text, turns away from textuality and against textology. It is the atextual texture. It is, in the end, the center *and* the beyond of the horizon.

What we wanted to show here—or suggest—is that in a certain way the *Timaeus* condenses what has always been in the works. The *Timaeus*'s relentless dynamic of the one and order unfolds at the same time as it holds within itself the inchoative of a reversal. With the precision of the demiurge, Plato seals off all of the breaches that he himself (with an astonishing audacity) opens up. Yet some thing (or some nonthing indeed) shakes this edifice at its core: (the) *khōra* deconstructs the imperturbable order that was announced (*khōra* is neither a word nor a concept: exactly like différ*a*nce!). *Khōra* cannot be avoided (Plato in all likelihood would have avoided formulating a superfluous hypothesis; he would have anticipated the Hegelian instructions concerning this and avoided resorting to any useless *muthos*, even if it were *eikos*). The consequence of this is considerable: The very condition of the possibility for the architectonic gesture also holds within itself the principle of its being revoked in the future. One can find the outline of a fissure at the heart of the logic of one of the principal texts that put an omniorder into place—an order that was balanced and begotten by the all-one. Hidden in the unsaids of (the) *khōra* lies the outline of this fissure that, *in fine*, could—would—ruin our uni-verse (as a unique and ordered *construction*).

What Derrida noted is worthwhile as an analysis of Plato's text and also as that which reveals or brings to light its internal vacillating and, well beyond this, its exercise—or practice—of struction. What Derrida *does* with the text is already outside of any con-struction. To isolate the philosophy-of-Plato within this nested entanglement "would be to misrecognize or violently deny the structure of the textual scene, to regard as resolved all of the questions of topology in general, including that of the places of rhetoric, and to think one understood what it means to receive, that is, to understand. It's a little early. As always."[75] The question is therefore to *defer* and *differentiate* again, always. The question is evidently not—Derrida agrees on this point quite explicitly—to forbid oneself from evoking the philosophy of Plato. To call Platonism a kind of abstraction is "neither arbitrary, nor illegitimate,"[76] in the same way that it is clearly not a question here of denying all that is *constructed* or banning any *allusion* to an order. This would not only be useless and artificial (and hence artifactual and therefore constructed!), but also impossible. Still, on the other hand, it is possible and even desirable to think a *praxis* of exiting, and to pose the question of a breach beyond or a sense outside of the constructivist order. "What I want to emphasize is simply that the passage beyond philosophy does not consist in turning the page of philosophy (which usually amounts to philosophizing badly), but in continuing to read philosophers *in a certain way*,"[77] Derrida writes in another context; what we wish to underscore is that thinking struction does not consist in renouncing all construction or ordering (which essentially amounts to deconstructing badly), but in pursuing composition/decomposition so as to position (oneself) *in multiple ways*. This "genre beyond genre" opened by (the) *khōra* could also serve as a guide or reference point for thinking the "sense beyond sense" that is imposed by struction.

V

The circle, the referral, the distorting mirror, the altered and alternating echo of the one and order that haunts Western thought since at least Plato is not content with governing over possible constructions. It also imposes the preconception of an organization or regularity. The circle of the one and order underlies and underpins what Derrida called a hauntology (an ontology of what is always already coming back to haunt us); it constitutes the irreducible—or rather the indivisible—assemblage, the *atomos* of *kosmos*. Still the impulse due to the one and order's own excess, the influx that results from its overcoming and reversal are as old as—so it seems—the putting into place and the putting into movement of the dialectic of the one and order.

To think struction—"the uncoordinated simultaneity of things or beings, the contingency of their belonging together, the dispersion of profusions of aspects, species, forces, forms, tensions, and intentions (instincts, drives, inclinations, and momentums)"[78]—is therefore not a mere deviation from the paradigm. It is both an inevitable consequence of what has always been known and a radical upheaval of all of the possible beings-*in*-the-world(s). Still, this difficult and necessary step [*pas*] can only be taken *precisely* by freeing oneself from the one and order. The deconstruction of logocentrism must be accompanied by a deconstruction of a cosmocentrism that is perhaps even more deeply buried in the sedimentary strata of our culture. The whole of the dialectic, and not simply one of its aspects or phases, must be halted. A lightening up [*une ralentie*, both a slowing down and a clearing in the sky] must win out over this back and forth—this taxonomic or reductionist—movement. It is not merely a territory to explore or a land to invent, but rather a putting into *dis*order that must be enacted.

Hilary Putnam, in the conclusion to "Irrealism and Deconstruction," a chapter in his synthesizing work *Renewing Philosophy*, sums up perfectly the way in which dominant thought, especially that of the analytic persuasion, tries to thwart this rupture: "[D]econstruction without reconstruction is irresponsibility."[79] The possibility of a structional prolonging, overcoming, and reversal of deconstruction is not even considered or considerable (it is outside of order and therefore outside of the world). And this is not only a question of coherence—Putnam proclaims very clearly—but also one of responsibility! In other words (and this is an old argument almost as automatic or machine-like as the modality it supports): nihilism is lurking. This also signifies: There is only the way, the trail, the path; there is no advancement in errancy. Thus here re-constructing is not only an aesthetic position; it is an ethical obligation. However, it seems to us that nihilism is rather—and more rigorously—the exhaustion of the system in its own schema, the decline of the model as this model unfolds, the crack in the canvas as it is being exhibited, the collapse of the passage under the weight of its own pillars. This is, therefore, the crumbling of the constructivist paradigm under its own weight, the collapse of a building that has literally become *unsupportable*. Gravity. The structional leap is clearly not insignificant. A lot must be reinvented with the thinking of disorder as our new starting point. We still do not know how to write or name disorder. How to designate or denote it. None of our idioms refer to it. All of our words are in the former world. One can write its name or form the grapheme of disorder's mark. But its concepts are held captive by the other schema. We have not even begun to envision the leap into disorder.

At a time when political action can essentially be summarized—even in its symbolism!—by the phrase *maintain order* and by exalting law enforcement [*forces de l'ordre*] at any price (and history has shown how exorbitant this price can be), one can easily imagine the societal importance of a (re) habilitation of an exit from the completely-normed, completely-settled, completely-regulated, completely-structured, and completely-aligned. An exit that is not thought of as a destitution or acquiescence, but as an abundance or ambition. This is not the place to explore such a path that, strictly speaking, is not even a path. Still, one could at least remark and deplore that this necessity, this "requirement of reason," writes Jean-Luc Nancy, which consists in "casting light on its own obscurity, not by bathing it in light, but by acquiring the art, the discipline, and the strength to let the obscure emit its own clarity" is evidently not current or even on the *order* of business [*à l'*ordre *du jour*].[80]

Thus, the question today would be "to recognize 'sense' in struction."[81] Indeed this would be a way of initiating the transition, of approaching the gulf without trembling, of coming to terms with our vertigo. But here again things are not so simple. Or perhaps, on the contrary, they are radically more simple: It may be possible—and even probable—to do without sense. To abstract oneself from it absolutely, to abandon its smallest fragments, to renounce its reassuring familiarity.

A first and fundamental displacement from signification to sense has already taken place: "[t]he in-significant is not that which is mean, without importance. It is the most important: the place where sense still detaches itself, disconnects itself from any signification."[82] Perhaps the overcoming of construction invites one to untie sense and allow sense itself to disintegrate, and also to consider a second movement, a new dislocation, an even more crucial removal.

The word—the vocable—sense is polysemantic even down to its etymology. This ambiguity or matrix-like plurivocity cannot and must not be reduced. "[T]he whole sense of sense is thus at least the unassignable unity of sensate sense and directional sense. . . . The unity of the senses of sense implies thus the original difference or heterogeneity of at least two senses."[83] This means—Nancy continues, anticipating the inevitable reproach, a reproach that has nothing to do with the proposition but will be uttered without fail (like a *conditioned* reflex)—"*neither* that we have to orient ourselves in a state of complete blindness, *nor* that it is a matter of indifference that we are disoriented, and that there is no difference between the best and the worst."[84] What this reflection on the *Sense of the World* has allowed one to understand and *sense* is in the end the consubstantiality or the co-ex-istence of sense and the world.

Still, if the world, therefore, and its share of one and order, must be addressed again and consequently abandoned ("There is no longer any world: no longer a *mundus*, a *cosmos*"),[85] perhaps the very idea of sense must be renewed. It is impossible to be lost without a compass (or be anything but lost without a compass, which amounts to saying the same thing) if there are no longer any poles. All of the paths are equal to one another on the surface of a sphere; all great circles are geodesic. Since the notion or intuition of sense can only be—at least partly—directional, and thus submitted to an orientation and a geo-metry, which here actually should be called a cosmo-metry, it is never entirely adjusted for the noncoordinated presence that struction calls for. *Efflorescence*, *Efflorese*nse: the light touch of essence and the efflorescence of sense. The base or core [*fond*], being, and nature withdraw themselves after barely being touched. Sense is dissolved in its own lushness. In what it claims to be its equivocity. Indeed, we are "not without an advancing, a roaming, a crossing."[86] Yet this scattered entanglement that does not allow itself to be surveyed except on the mode of a "destinerrance" can no longer be thought within the regime of sense. Destinerrance is anything but remaining-in-place: It is a retroactive comprehension (the recipient, one could say, is the one who writes the message),[87] which through this reversal troubles sense, spreads it, and dilutes it. And, necessarily, neutralizes it. Neuter: once again, a third gender beyond gender. If there is a sense or a possibility of sense, then a "common sense" is always appearing on the horizon (the common sense of Kant, the one that settles or founds a demonstration). And it imposes its direction, orientation, and governance. If, on the other hand, all of the senses make sense, it is sense itself that is senseless. One must perhaps come to terms with it and think outside of sense—in other words, to think in *any and every* sense. Demonstration must become for monstration what deconstruction is for construction.

We do not yet know what the thought of—and in—struction may become, but surely we can already affirm that thought is not extinguished within it. If only because this new landscape (this acosmic space) is created and fashioned—*structed*—by thought. If there is a danger here—let us insist one last time on this crucial point—it is certainly not that of nihilism, which here is a contradiction in terms or oxymoron, even if the logic that is contravened is no longer that of *logos*. What must be broken away from at long last is the old, conformist, and safe response according to which the not-yet imagined or (re)presented can only be a depleted void or a figure of nothingness.

There is clearly something (or some post-thing) outside of oneself. This relation-to still has to be thought out, but in a movement or gesture that is a sort of anti-cosmology.[88] It is neither *cosmos* nor *logos*. It is only an "archi-writing" of chaos (or *cosmoi*, as long as the plural is not subsumed into a new order). Nothing is more artificial and unnecessarily abstruse than a systematic and parallel comparison of the great ancient cosmogonies and the contemporary model of astrophysics. Nevertheless, one point is striking here: the role of Khaos in the former is extraordinarily close to that of the Big Bang in the latter. Both Khaos and the Big Bang are only mentioned in a highly allusive way. They have to *be*, but they cannot be thought. They remain in the distance, asymptotically far away. They are principles more than beings, and yet they are also aporias more than solutions. Not one line is devoted to Khaos in the Hesiodic theogony, and generally speaking just as little is devoted to the Big Bang (as such) in the treatises on physical cosmology. Rift or fault in the former, and singularity in the latter. Abyss on the side of myth, and divergence on the side of the sciences. Much later Aristophanes will try to assimilate chaos with the air and physicists will try to reduce the Big Bang to the end of the inflationary phase or to replace it with a bounce. Which one is of little importance. A gaping void cracks open and a shaking or trembling imposes itself at the same time or by the same measure that thought goes back to this original strangeness. It is both impossible and necessary.

Clearly, today no one can foresee what a philosophy of this new paradigm will be or could be. The language of the "all together," of the non-world, or of what happens after the leap has not yet been invented. One can presuppose nevertheless that one will have to reread or in this case read *for the first time* (what, in short, Pessoa never stopped suggesting)[89] Nietzsche and William James, about whom Rorty wrote: "Occasional protests against this conception of culture as in need of 'grounding' and against the pretensions of a theory of knowledge to perform this task (in, for example, Nietzsche and William James) went largely unheard."[90] This certainly does not imply a grand book burning of the philosophical tradition, but a way of using the philosophical tradition in order to benefit from the gift of its careful analysis of all of the impossible *puttings-into-order*. In short, as always: to turn construction back against itself. Not for destruction's sake but for actualizing all of the assemblages that thus far have only been virtual. For enduring all of the dispersions that thus far have been neither admissible nor conceivable. One might subsequently have to consider, in another context of course, Howard Zinn, as he underscores that the role of history does not consist at all in bringing to light an inevitable, causal chain of events, but, on the contrary, in showing that "things could have happened otherwise": Philosophy will be able to be (re)read as a systematic study of taxonomies that have been worn out by the repeated frictions that were caused by their own rigidity. Far from being useless, tradition will demarcate the unsurpassed limits so as to allow one to enter into the "other" side. Where nothing can be constructed anymore. In the nonplace of a dis-united dis-order.

((Beyond—that is, beneath—construction, destruction, and deconstruction, the real becomes veiled. Not so much in the sense that Bernard d'Espagnat gives to the term "veil" as an exegesis of quantum mechanics,[91] but more in the infinitely ambiguous sense that Hélène Cixous and Jacques Derrida played with.[92] *Voile* [veil, *m.*; sail, *f.*]—*both* in the sense of the masculine *and* in the feminine. Or in the masculine *and then* in the feminine, at the whim of the elytron floating between the one and the other as Derrida said. It is uncertain or mixed. Outside of any place where genders are sedimented and (de)composed. Once again. Veil: It is what prevents one from observing-dissecting the nude *and* what propels, carries, and pushes. It is a regime of androgyny or a becoming hermaphrodite. A veil so finely woven that the work of warping—the work of putting into order—is concealed, faded, masked: a fusion unto continuum "within the textual, the textile, and the histological."[93] Less the weaver of the *Republic* and more the weaver of the *Phaedrus* (in Derrida). A shaded and windswept real. But one that is also veiled like something worn down or deteriorated, or, as one says [in French], like a *veiled* [bowed or warped] piece of wood. Uneven or distorted. The loss of the plane, flatness, the flat field. A veil that casts a shadow over the gaze and blurs vision. The real of the myopic person: unfocused, and, in fact, outsourced [*délocalisé*]. An unseen utopia. A myopia, whose benefits are suggested by Hélène Cixous—or, more precisely, that Cixous underscores the risks of curing.[94] The risk of the exhibited, shown, and demonstrated real—a real that is more obscene than it is indecent. And atrophied by this false *coup de* gift [*coup de don*].[95] "But should that veil be suspended, or even fall a bit differently, there would no longer be any truth, only 'truth'," wrote Derrida as he spurred Nietzsche (or let himself be spurred by Nietzsche).[96] A complicity, then, between the naked, offered, and unbashful truth and its radical absence. Further on, Derrida evokes "the veil of truth and the simulacrum of castration"—not *on* truth but *of* truth.[97] Is it a veiled truth or a veil becoming truth? It is undecidable, naturally. It is a veil *applied* to the truth and *due* to the truth. A veil that has become true in its capacity to intricately intertwine the truthful. Truth that has become uncertain because of its propensity to fold veils, to pack up its veils and hoist its sails [*plier voile*].[98]

A veil either veils or sails off. But a veil also has value by itself. In the way it falls. Drapes. Hangs. A veil combines and mixes threads into the weft. A veil abstracts itself from ties. What remains are the folds, the tails. The limit of the *dévoi(l)ement*, the *unveiling* and the *straying from the path*)).

The question is no longer to make sense, but to leap. Or to listen: "[T]o be listening is always to be on the edge of meaning, or in an edgy meaning of extremity."[99] To overthrow the dialectic of the one and order, to let it get caught up in the game of its own exhaustion and leave it there. To let the construction and deconstruction of cosmo-teleo-centrism be transformed into the struction of chaologism. To be immersed in what Bruno Pinchard calls a "polyphony of disorder."[100] It is the question of a revolution that would destabilize the entire work or tradition. And through its disintegration, what would open up is a continuous infinity of possibles. It is not a matter of thinking differently, but of thinking in a place where thought does not have any currency [*avoir cours*]. And finally, it is a question of unistruction: the struction of the ones, a contingent and proliferating struction, one without order and one that does not subsume singularities.

A. Barrau

NOTES

FOREWORD

I would like to thank my colleague Chelsea C. Harry, assistant professor of philosophy at Southern Connecticut State University, for her generous assistance with the final preparation of this "Foreword."

1. Jean-Luc Nancy, *The Inoperative Community*, trans. Peter Connor (Minneapolis: University of Minnesota Press, 1991). Hereafter cited as *IC* followed by the page number. An early, partial version of *La Communauté désoeuvrée* (Paris: Christian Bourgois, 1986) appeared as an article in the journal *Aléa*, no. 4 in 1983.

2. Aurélien Barrau and Jean-Luc Nancy, *What's These Worlds Coming To?* trans. Travis Holloway and Flor Méchain (New York: Fordham University Press, 2015). Hereafter cited as *WTW* followed by the page number.

3. See, particularly §26 of Martin Heidegger, *Being and Time*, trans. John Macquarrie and Edward Robinson (New York: Harper and Row, 1962). Hereafter cited as *BT* followed by the page number. Heidegger writes, for example, "Dasein in itself is essentially Being-with" (*BT* 156).

4. Jean-Luc Nancy, *Being Singular Plural*, trans. Robert Richardson and Anne O'Byrne (Stanford, Calif.: Stanford University Press, 2000), 83. Hereafter cited as *BSP* followed by the page number.

5. Jean-Luc Nancy, *Corpus*, trans. Richard Rand (New York: Fordham University Press, 2008), 19. Hereafter cited as *C*, followed by the page number.

6. For a treatment of the ethical dimensions, or the ethicality of the coexposition of *being-with*, see François Raffoul, *The Origins of Responsibility* (Bloomington: Indiana University Press, 2010). Hereafter cited as *OR*, followed by the page number. See, for example, pp. 15, 206, 221–24, and 260. On page 15 of *OR*, Raffoul writes that "responsibility exceeds the anthropocentric closure, and is to be situated in the between of singularities." Further, Raffoul writes, "The 'with' is coextensive with being, *so that the ethical is*

co-extensive with the ontological. Nancy emphasizes this point in *Being Singular Plural* by insisting on the indissociability of being and the other" (*OR* 206).

7. Jean-Luc Nancy, *The Sense of the World*, trans. Jeffrey Librett (Minneapolis: University of Minnesota Press, 1997), 3. Hereafter cited as *SW* followed by the page number.

8. We can note that while the French term "*l'immonde*" refers to something that is vile or disgusting, in the context of the work of Jean-Luc Nancy regarding *le monde* [world], *l'im*monde can take on the connotation of an un-world, a world that has undergone the process of globalization, in contrast to an authentic world forming. See the translation of "*l'immonde*" as "unworld" on page 34 of *The Creation of the World or Globalization*, trans. François Raffoul and David Pettigrew (Albany: State University of New York, 2007). Hereafter cited as *CW* followed by the page number.

9. Martin Heidegger, *Bremen and Freiburg Lectures: Insight Into That Which Is and Basic Principles of Thinking*, trans. Andrew J. Mitchell (Bloomington: Indiana University Press, 2012). Hereafter cited as *BL* followed by the page number.

10. Nancy addresses the crisis of a human existence that is ruled by a "general equivalence" within a logic of economic profit in his text *L'Équivalence des catastrophes: (Après Fukushima)* (Paris: Editions Galilée, 2012). He characterizes this phenomenon of general equivalence as catastrophic, and, in particular, as a catastrophe of sense" (20, my translation).

11. Jean-Luc Nancy, *La possibilité d'un monde: Dialogue avec Pierre-Philippe Jandin* (Paris: Les petits Platons, 2013), 34 (my translation). Hereafter cited as *PM* followed by the page number.

12. Jean-Luc Nancy, "Heidegger's 'Originary Ethics,'" in *Heidegger and Practical Philosophy*, ed. François Raffoul and David Pettigrew (Albany: State University of New York Press, 2002), 65–85. Hereafter cited as *HPP* followed by the page number.

13. Jean-Luc Nancy, "The Being-with of the Being-There," trans. François Raffoul and David Pettigrew, in *Rethinking Facticity*, ed. François Raffoul and Eric Nelson (Albany: State University of New York Press, 2008), 113–27. Hereafter cited as *RF* followed by the page number.

14. See Jacques Derrida and Maurizio Ferraris, *A Taste for the Secret*, trans. Giacomo Donis (Cambridge: Blackwell, 2001). Hereafter cited as *TS*, followed by the page number. In his interview with Maurizio Ferraris, Derrida thematizes a certain excess with respect to meaning or sense. He states that in *writing*, there is a certain paradoxical hope that one will not be understood immediately or completely, because "a transparency of intelligibility would destroy the text, it would show that the text has no future [*avenir*], that it does not overflow the present, that it is consumed immediately" (*TS* 30). Rather, Derrida seeks a "certain zone . . . of not-understanding," a zone that would provide "a chance

for excess to have a future . . . and to engender new contexts" (ibid.). If something is "totally intelligible," and is "totally saturated by sense," he states, "it is not given to the other to read" (*TS* 31). This chance for *excess*, then, a genuinely hospitable "*giving to be read*," leaves "the other room for an intervention by which she will be able to write her own interpretation" (ibid.). Derrida asserts that he has "a taste for the secret," because of his "fear or terror in the face of a political space . . . that makes no room for the secret" (*TS* 59). For Derrida, "if a right to the secret is not maintained, we are in a totalitarian space" (ibid.).

FRONTMATTER

1. Claude Lévi-Strauss, *Anthropology Confronts the Problems of the Modern World*, trans. Jane Marie Todd (Cambridge, Mass.: Harvard University Press, 2013), 86–87.

PREAMBLE

1. [Although the terms *globalisation* and *mondialisation* are used interchangeably in French, Nancy discusses a distinction between these terms in *The Creation of the World or Globalization*. For further reference, see Jean-Luc Nancy, *The Creation of the World or Globalization*, trans. François Raffoul and David Pettigrew (Albany: State University of New York Press, 2007), 1–26.—Trans.]

2. [These two expressions are pronounced the same way in French despite being written and read differently. The phrase without parentheses is a common idiomatic expression in French. The addition of the letter *s* in parentheses indicates the authors' alteration of the expression in which case the noun "world" is made plural. Please see the "Translators' Preface" for further reference.—Trans.]

MORE THAN ONE

The initial version of this text was written for a symposium devoted to Derrida in November 2009 in Madrid, organized under the direction of Cristina de Peretti (UNED).

1. Pico della Mirandola, "On the Dignity of Man," in *Pico Della Mirandola*, trans. Charles Glenn Wallis (Indianapolis, Ind.: Bobbs-Merrill, 1940), 10.

2. Jacques Derrida, *Writing and Difference*, trans. Alan Bass (Chicago: University of Chicago Press, 1978), 299.

3. [*Rien*, the French word for "nothing," is derived from the Latin *res*, *rem*, meaning "thing."—Trans.]

4. The only one of our contemporaries who has intended to build an "ontology" in proper terms—Alain Badiou—calls it an "ontology of the

multiple" and puts it under the sign of the proposition "the One is not [*pas*]." He thus both takes up and interrupts this history at the same time by opposing the non-Being of the event to the not-Being [*ne-pas-être*] of the One. In Derrida, the (un)doubling of the origin opens up the evental in Being, or rather, in the space left open by the Heideggerian *Destruktion* of every ontology. The one does not have to be or not to be if Being itself is not [*pas*]. Still it occurs as a rhythm. In other words, it *is itself arriving* to itself, it is coming about [*survenir*] to itself—as "more than one." It could be noted that an asymptote is common to these two ways of proceeding—according to which the one would be its own excess. In a manner analogous to the *Ereignis* of Heidegger, on the one hand (which is the "appropriation" of the non-one of the own onto the "unicity" of Being—*Einzigkeit des Seyns*, which appears, for example, in Heidegger's *Beiträge*, §12), and on the other hand, to the "univocity of Being" of Deleuze (which is only made out of different/ciations). The sort of web I am therefore hastily weaving does not aim to weave together all these thoughts into a net, but rather into the perspective of a question such as this one: What, therefore, is *an* epoch? Where does the unity of a moment, a stage, or a leveling play out? There never is a pure heterogeneous disparity between the thoughts of an epoch, even though we know little about what constitutes it as an "epoch" (and therefore a "one") and even if there are at the same time noticeable dissonances. "An epoch" is not a consensus, but a relative possibility to refer the ones back to the others even down to the sharpest dissensus.

5. [The expression "*sens unique*" (literally, "unique sense") appears on road signs in France to signal one-way streets.—Trans.]

6. [Nancy breaks the rules of grammar by writing "*pluralité des 'un'*," in which case he removes the *s* from the expected plural *uns*. We have indicated his choice with the English "one" rather than the grammatically correct plural "ones."—Trans.]

7. [Nancy is differentiating between two uses of the word *plus* in French: one whose final *s* is pronounced, which is translated into English as "plus" (as in addition), and another in which case *plus* has a final *s* that is silent, which is translated into English as "more than."—Trans.]

8. To concatenate some quaternary with some ternary in Derrida, it should be considered how he receives the *Geviert*, about which he does not fail to recall that it is formed, for Heidegger, "from one *original* Unity" (cf., for example, Jacques Derrida, *Dissemination*, trans. Barbara Johnson (Chicago: University of Chicago Press, 1981), 354.

9. André Breton, *Communicating Vessels*, trans. Mary Ann Caws and Geoffrey T. Harris (Lincoln: University of Nebraska Press, 1990), 134.

10. Pico cites Anselm: "*Domine, non solum es quo majus cogitari nequit; sed es quiddam majus, quam cogitari possit*" (*Proslogion*, Ch. XV) [Pico writes: "It can

be concluded from this that God is not only that than which no greater can be conceived, as Anselm said, but he is that which is infinitely greater than every thing that can be thought," Pico della Mirandola, "On Being and the One," in *Pico Della Mirandola*, trans. Wallis, 53.—Trans.]

11. *De Ente et Uno*, cited from the edition of the text given by André-Jean Festugière in the *Archives d'histoire doctrinale et littéraire du Moyen-Âge*, vol. 7 (Paris: Vrin, 1932).

12. Thomas Aquinas, *Summa Theologica*, 1a.11.

13. Today the position of science is turning away from unification: "Would it not be better to renounce the proliferation of the worlds and the evocation of correct versions, each constituting a world, in order to return to the more orthodox image of different descriptions of a one and only, neutral and subjacent world? Yet the world thus recovered, Goodman writes, 'has no genre or order or movement or rest or structure—it is a world that does not merit one's struggle for or against it.' It is one hypothesis too many. This 'passion of the one,' as Goodman calls it, of the subjacent, of the fundamental, of Being or of the *archē*, atrophies the possibilities. It is not necessarily subsumed under an overarching concept. It also closes pathways and irrigation canals. It prunes the branches and damages the small roots." Aurélien Barrau, "La cosmologie comme manière de faire un monde," in *Forme et origine de l'Univers*, ed. Aurélien Barrau and Daniel Parrochia (Paris: Dunod-La Recherche, 2010), 371.

14. Maurice Blanchot, *The Writing of the Disaster*, trans. Ann Smock (Lincoln: University of Nebraska Press, 1995), 138.

15. [A term for currency that is no longer commonly used; from the Latin *numerarius* (pertaining to number, calculation).—Trans.]

LESS THAN ONE, THEN

1. Victor Hugo, *Les Misérables*, Trans. Charles E. Wilbour (New York: The Modern Library, 1992), 458.

2. Jacques Derrida, *Glas*, trans. John P. Leavy Jr. and Richard Rand (Lincoln: University of Nebraska Press, 1986).

3. "Thought weighs precisely the weight of sense." Jean-Luc Nancy, *Le Poids d'une pensée, l'approche* (Strasbourg: La Phocide, 2008). 12.

4. In other words, its potential "disassembling of enclosed bowers, enclosures, fences." Jean-Luc Nancy, *Dis-Enclosure: The Deconstruction of Christianity*, trans. Bettina Bergo, Gabriel Malenfant, and Michael B. Smith (New York: Fordham University Press, 2008), 161.

5. Cf. Charles Ramond, "Derrida: Hegel dans Glas," lecture given at the symposium "Les interprétations de Hegel au XXè siècle," Poitiers, September 29, 2006.

6. A hieroglyphic writing but also a drawing or design (see Jean-Luc Nancy, *Le Plaisir au dessin* [Paris: Galilée, 2009]): "Egyptian figures that go forth always at the same pace with their heads turned backwards and their gaze always fixed on their invisible provenance," writes Philippe Lacoue-Labarthe concerning literature in *Phrase* (Paris: Christian Bourgois, 2000), 23.

7. Cf. Jean-Luc Nancy, *The Sense of the World*, trans. Jeffrey S. Librett (Minneapolis: University of Minnesota Press, 1997).

8. Cf., for example, Jacques Derrida, "A Silkworm of One's Own," in Hélène Cixous and Jacques Derrida, *Veils*, trans. Geoffrey Bennington (Stanford, Calif.: Stanford University Press, 2001).

9. [Barrau coins a hyphenated neologism, *un-version*, which is pronounced in the same way as the French word *inversion*. Although *un-* is pronounced the same as the common French prefix *in-*, *un-* itself is not a common prefix in French. Since here the word is already placed next to *inversion*, and since textually *un-* indicates simply "one," *un-* is translated as "uni-." For further context on the use of this prefix, see the concluding paragraph on p. 88—Trans.]

10. Cf., for example, the preface by Jean-Louis Backès to Hesiod in *Théogonie. Les Travaux et les Jours* (Paris: Gallimard, 2001).

11. This is demonstrated by Jean-Paul Dumont, for example, in his introduction to *Les Présocratiques* (Paris: Gallimard, 1988).

12. David Lewis considers that a counterfactual assertion, such as, "If Aristotle had not restored *mimesis*, the history of art would have been entirely different," makes sense because, indeed, a world exists in which Aristotle has not written his works and in which the development of the arts has been affected by it.

13. Nelson Goodman, *Ways of Worldmaking* (Indianapolis, Ind.: Hackett, 1978), x.

14. Irrational numbers cannot be written in the form of [finite] fractions. Transcendental numbers are not even solutions of polynomial equations with rational coefficients. For example, the infinite summation of the reciprocals of the squares is indeed a transcendental number despite the fact that each element of the sum is rational.

15. One could add, echoing the work of Boghossian, that it is also the opposite of a "fear of knowledge."

16. To dismember: "to cause the disappearance of a thing as a unit" (Rutebeuf, *La Discorde de l'Université et des Jacobins*, 35, in *Oeuvres complètes*, vol. 1, ed. Edmond Faral [Paris: Picard, 1969], 240).

17. J. L. Austin, *How to Do Things with Words* (New York: Oxford University Press, 1962).

18. Hilary Putnam, *Renewing Philosophy* (Cambridge, Mass.: Harvard University Press, 1992).

19. Gilles Deleuze, *Difference and Repetition*, trans. Paul Patton (New York: Columbia University Press, 1994).

20. Inconsistent precisely because the relationship between these two repetitions is repetition itself. Cf. Jacques Derrida, "Plato's Pharmacy," in *Dissemination*, trans. Barbara Johnson (Chicago: University of Chicago Press, 1981), 61–171.

21. Nelson Goodman, *Languages of Art* (Indianapolis, Ind.: Hackett, 1976).

22. Richard Rorty, *Objectivity, Relativism, and Truth* (Cambridge: Cambridge University Press, 1991).

23. Jacques Derrida, *Margins of Philosophy*, trans. Alan Bass (Chicago: University of Chicago Press, 1982).

24. Jean-Michel Salanskis "Nouvelles frontières," in *Le Magazine Littéraire*, no. 498, "Derrida en héritage," June 2010, 68.

25. Bruno Latour and Isabelle Stengers show the Deleuzian and pragmatist resonances within this expression in their introduction to Étienne Souriau, *Les Différents Modes d'existence* (Paris: PUF, 2009).

26. Jacques Derrida, *The Truth in Painting*, trans. Geoff Bennington and Ian McLeod (Chicago: University of Chicago Press, 1987), 2.

27. See Goodman, *Ways of Worldmaking*. This position is frequently judged as *subversive* because it dissolves the border between art and science. Borders, as we know, are reassuring.

28. Derrida, *The Truth in Painting*, 2.

29. See Goodman, *Languages of Art*.

30. Jean-Luc Nancy, *The Muses*, trans. Peggy Kamuf (Stanford, Calif.: Stanford University Press, 1996), 18.

31. Theodor Adorno, *Aesthetic Theory*, ed. Gretel Adorno and Rolf Tiedemann, trans. Robert Hullot-Kentor (Minneapolis: University of Minnesota Press, 1997).

32. Vladimir Jankélévitch, *Music and the Ineffable*, trans. Carolyn Abbate (Princeton, N.J.: Princeton University, 2003), 18.

33. We are not referring here to the trend that was started by Tristan Murail and Gérard Grisey, even though their understanding of spectrality (i.e., literally, of music that is *spectered*) in the sense of exiting series or hegemonic structures is extremely well suited to our topic.

34. Surely among the most sober and profound commentaries on the famously perilous subject of the exegesis of Bach's major works are those of Gilles Cantagrel.

35. This also suggests that fugues should never end, be finished, or be completed. They should not have an untangling, a closure, or a conclusion. They are infinite in essence.

36. Cf. Gilles Deleuze, *Nietzsche and Philosophy* (New York: Columbia University Press, 2006), 147. More generally, it is the Deleuzian idea of eternal return as "one of the multiple" that is important here.

37. Jacques Derrida, *Writing and Difference*, trans. Alan Bass (Chicago: University of Chicago Press, 1978), 248–49.

38. "We should get rid of our superstitious valuation of texts and written poetry. Written poetry is worth reading once, and then should be destroyed," writes Artaud! Derrida cites this quotation in *Writing and Difference*, 247.

39. Supersymmetry has around five times more free parameters than the "standard model" of elementary particles.

40. Cf., for example, Aurélien Barrau, Patrick Gyger, Max Kistler, and Philippe Uzan, *Multivers* (Paris: La Ville Brûle, 2010).

41. Of course, there exist attempts to unify these interactions. The simplest and closest model (founded on the symmetry group SU[5]) to the one successfully developed for subsuming the electromagnetic and weak nuclear forces (founded on SU[3]XSU[2]XU[1]) is excluded nevertheless.

42. This division—which is correct from an operational point of view—is not rigorously precise. Yet quantum mechanics and general relativity indeed have their own chosen fields that are essentially mutually exclusive.

43. There are many good, introductory works to these models. See, for example: Jean-Pierre Luminet, *Le Destin de l'univers* (Paris: Fayard, 2006); Carlo Rovelli, *What Is Time? What Is Space?* trans. J.C. van den Berg (Rome: Di Renzo Editore, 2006); Alain Connes, *Noncommutative Geometry*, ed. Marc A. Rieffel, trans. Sterling Berberian (London: Academic Press, 1994). Here we are interested simply in highlighting the diversity of these works.

44. Jacques Derrida, *Positions*, trans. Alan Bass (New York: Continuum, 2002), 105. In this respect, Derrida delivers a message on truth that is deliberately *undecidable*—or as precise as possible, in other words: "[I]t goes without saying that in no case is it a question of a *discourse against truth* or against science. . . . I repeat, then, leaving all their disseminating powers to the proposition and the form of the verb: we *must have* [il faut] truth."

45. There are three breaches then: (1) a multiplicity of phenomena, objects, and laws; (2) a multiplication of "independent" disciplinary fields at the very core of theoretical physics; and (3) a multitude of different and correct theories (at any given moment) that confront the same problem (let us give the same sense to this word that Deleuze did—that of an insistence of Being; cf. Véronique Bergen, "Deleuze et la question de l'ontologie," *Symposium* 10, no. 1 (2006): 7–24).

46. Deleuze and Guattari were precursors, almost visionaries, in establishing that science is "haunted not by its own unity" but by its relationship to chaos. See Gilles Deleuze and Felix Guattari, *What Is Philosophy*, trans. Hugh Tomlinson and Graham Burchell (New York: Columbia University Press,

1994), 42. This relationship is obviously not without complicities and sympathies with the enemy. We will return to this point.

47. This is because indeed what characterizes the stanzas or *stances* (*stanza*: dwelling [*demeure*]) is the completeness of the sense that is associated with each iteration and the rhythm that they give to the entire work: These are exactly the two fundamental characteristics we are referring to here.

48. "So if anything at all happens to writing, nothing happens to it but *touch*." Jean-Luc Nancy, *Corpus*, trans. Richard A. Rand (New York: Fordham University Press, 2008), 11. Touch reveals itself as "proximate distance" (Nancy, *The Muses*, 17). For a detailed study on the importance and reoccurrence of touch in Nancy, one can naturally refer to the work that Derrida dedicated to Nancy: Jacques Derrida, *On Touching—Jean-Luc Nancy*, trans. Christine Irizarry (Stanford, Calif.: Stanford University Press, 2005).

49. Jean-François Lyotard, *The Postmodern Condition: A Report on Knowledge*, trans. Geoff Bennington and Brian Massumi (Minneapolis: University of Minnesota Press, 1984), 64.

50. Ibid., 66.

51. Ibid., xxv.

52. "[F]or as long as the world was essentially in relation to some other (that is, another world or an author of the world), it could *have* a sense. But the end of the world is that there is no longer this essential relation, and that there is no longer essentially (that is, existentially) anything but the world 'itself.'" Nancy, *The Sense of the World*, 8.

53. "I try to keep myself at the *limit* of philosophical discourse. I say limit and not death, for I do not at all believe in what today is so easily called the death of philosophy." Derrida, *Positions*, 6.

54. Cf. Michael Serres, *The Birth of Physics* [*La Naissance de la physique dans le texte de Lucrèce*], trans. Jack Hawkes (Manchester, U.K.: Clinamen Press, 2001).

55. "Now, the 'thought-that-means-nothing' . . . is given precisely as the thought for which there is no 'sure opposition between outside and inside'" (Derrida, *Positions*, 12).

56. Ibid.

57. This idea can be found, for example, in the course of a reflection on structuralism: "that the center had no natural site, that it was not a fixed locus but a function" (Derrida, *Writing and Difference*, 280).

58. Jacques Derrida, *On the Name*, ed. Thomas Dutoit, trans. David Wood, John P. Leavey Jr., and Ian McLeod (Stanford, Calif.: Stanford University Press, 1995), 8.

59. Cf., for example, Anne-Gabrièle Wersinger and Sylvie Perceau, "L'auto-réfutation du Sceptique vue de la scène antique," *Revue de métaphysique et morale*, no. 65 (2010): 25–43.

60. For an exhaustive review, cf. Maria Baghramian, *Relativism* (London: Routledge, 2004), 136–41.

61. Kenneth Taylor, "How to Be A Relativist," a contribution to the debate *The Contingency of Facts and the Objectivity of Values*, available on Stanford University's prepublications server at http://www.stanford.edu/~ktaylor/relativism.pdf.

62. Cf., for example, Paul Boghossian, *Fear of Knowledge: Against Relativism and Constructivism* (Oxford: Oxford University Press, 2006), or Timothy Chappell, "Does Protagoras Refute Himself?" *Classical Quarterly* 45, no. 2 (1995): 333–38.

OF STRUCTION

An earlier version of this text was first published in *Die technologische Bedingung. Beiträge zur Beschreibung der technischen Welt*, ed. Erich Hörl (Berlin: Suhrkamp, 2011).

1. [The polysemantic word *la technique*, which is translated as "technology" in this essay, could also be correctly translated as "technique" or "technics." While in French *la technique* may suggest, like Aristotle's *technē*, a kind of skill, know-how, or technique, *la technique* may also suggest technology in the sense of Heidegger's *die Technik*, which has typically been translated into French as *la technique* and into English as "technology" or, more recently, as "technics." Whereas Heidegger generally does not translate *technē* with *die Technik* because he considers modern technology or *die Technik* as being very different from the Greek sense of *technē*, here Nancy views modern technology as a one of many "maturations" of *technē*. Elsewhere Nancy has even offered *la technique* as a translation for *technē*, for example, in Jean-Luc Nancy, "The Technique of the Present: On On Kawara," in *Multiple Arts: The Muses II*, ed. Simon Sparks (Stanford, Calif.: Stanford University Press, 2006).—Trans.]

2. Letter from Mallarmé to Eugène Lefébure on May 27, 1867: "I've created my work only by *elimination*, and any truth I acquired resulted uniquely from the loss of an impression which, having sparkled, burnt itself out and allowed me, thanks to the shadows thus created, to advance more deeply in the sensation of the absolute shadows. Destruction was my Beatrice . . . the sinful and hasty road, a road which is satanic and *facile*, the road of self-destruction which has produced not strength but a sensibility." Stéphane Mallarmé, *Selected Letters of Stéphane Mallarmé*, ed. and trans. Rosemary Lloyd (Chicago: University of Chicago Press, 1988), 77–78.

3. In "Tale," from Arthur Rimbaud's *Illuminations*, which can be found in Arthur Rimbaud, *Illuminations*, trans. John Ashbery (New York: Norton, 2011), 35. And one may also think of Dostoevsky: "Man loves to construct and lay down roads, no question about it. But why is he so passionately fond of

destruction and chaos? Tell me that! . . . Isn't man perhaps so passionately fond of destruction and chaos (and there's no disputing that he's sometimes very fond of them, that really is the case) that he himself instinctively fears achieving his goal and completing the building in course of erection?" Fyodor Dostoevsky, *Notes from Underground* and *The Double*, trans. Ronald Wilks (New York: Penguin Books, 2009), 30.

4. It so happens that *struction* is also a concept in graph theory, which is not relevant here.

5. Aurélien Barrau, "Quelques éléments de physique et de philosophie des multivers," *Laboratoire de Physique Subatomique et de Cosmologie CNRS-IN2P3*, http://lpsc.in2p3.fr/barrau/aurelien/multivers_lpsc.pdf, 122 (accessed September 27, 2013).

6. On this topic, see the use of the term "construction" in the work cited in the previous footnote.

7. Thinking about this on a simple level, one knows that a particle accelerator or a space probe is not independent from the "objects" it examines, and this is also conversely the case. But in truth we are only at the beginning: the intricate connection or involvement of the observer in observed reality, such that this reality never ceases to be amplified and made more complex in the so-called hard sciences as well as in the sciences that are called human, signifies in reality a progressive transformation of the status of "science." Even to speak of this "intricate connection" still suggests an implied agreement with a model of noninvolvement and "objectivity." Here as well, whereas it was once customary to think of technologies as applications of certain scientific results, today technology gives science an unprecedented status and unprecedented content.

8. For the Moderns, intelligence has a tendency to get confused with technology. This is why "artificial intelligence" (a tautology perhaps?) seems so fascinating. On the other hand, when one speaks in French of emotional intelligence as the "intelligence of the heart," one clearly indicates that one is using a metaphor.

. . . AND OF UNISTRUCTION

1. Cited by Jacques Derrida in "Khōra," in *On the Name*, ed. Thomas Dutoit, trans. David Wood, John P. Leavey Jr., and Ian McLeod (Stanford, Calif.: Stanford University Press, 1995), 148; translation here is that of David Farrell Krell in Martin Heidegger, *Nietzsche, Vol. I and II: The Will to Power as Art, The Eternal Recurrence of the Same*, trans. David Farrell Krell (New York: HarperCollins, 1991), 91–92.

2. [In the title of this chapter, Barrau invents a neologism, *unstruction*, which is pronounced in the same way as the French word *instruction*, about which

Nancy has written in Chapter 3. Although *un-* is pronounced the same as the common French prefix *in-*, *un-* itself is not a typical prefix in French. Since textually *un-* indicates simply "one," *un-* is translated as "uni-." Barrau will explain in his concluding paragraph that by "unistruction" he means "the struction of the ones."—Trans.]

3. Myth of order in the sense in which Marcel Detienne writes: "nothing is more concrete, more real, more obvious than myth," in *The Creation of Mythology*, trans. Margaret Cook (Chicago: University of Chicago Press, 1986), 130.

4. This is demonstrated by Brice Halimi.

5. Gottfried Wilhelm Leibniz, "Discourse on Metaphysics," in *Discourse on Metaphysics and Other Writings*, trans. R. Niall D. Martin and Stuart Brown (Manchester and New York: Manchester University Press, 1988), 43–44.

6. Richard Rorty, *Philosophy and the Mirror of Nature* (Princeton, N.J.: Princeton University Press, 1979), 3.

7. This approach is essentially motivated by field theory. From the viewpoint of general relativity, the presence of a cosmological constant explains this acceleration without difficulty and is perfectly justified from a theoretical point of view.

8. Jean-Luc Nancy, *The Muses*, trans. Peggy Kamuf (Stanford, Calif.: Stanford University Press, 1996), 3.

9. Indeed, Nietzsche had already written: "The *fundamental prejudice* is, though, that it is inherent to the *true being* of things to be ordered, easy to survey, systematic; conversely, that disorder, chaos, the unpredictable can only make its appearance in a world that is false or incompletely known—in short, that is an error—which is a moral prejudice, drawn from the fact that the truthful, reliable human being is a man of order, of maxims, and all in all tends to be something predictable and pedantic. And yet it cannot be demonstrated at all that the in-themselves of things follows this recipe for the model civil servant," in Friedrich Nietzsche, *Nietzsche: Writings from the Late Notebooks*, ed. Rüdiger Bittner, trans. Kate Sturge (Cambridge: Cambridge University Press, 2003), 42. Nietzsche took the first step. He was already beyond the thought of chaos or the conceptualization of a beyond-order and had entered the exercise or practice of a disordered thought. This is what we must work toward.

10. Gilles Deleuze and Félix Guattari, "Conclusion: From Chaos to the Brain," in *What Is Philosophy?* trans. Hugh Tomlinson and Graham Burchell (New York: Columbia University Press, 1994), 201–18.

11. Ibid., 208.

12. François Zourabichvili, "Plane of Immanence (and Chaos)," in *Deleuze: A Philosophy of the Event* together with *The Vocabulary of Deleuze*, ed. Gregg Lambert and Daniel W. Smith, trans. Kieran Aarons (Edinburgh: Edinburgh University Press, 2012), 188–99.

13. Gilles Deleuze and Félix Guattari, *A Thousand Plateaus: Capitalism and Schizophrenia*, trans. Brian Massumi (Minneapolis: University of Minnesota Press, 2004), 266.

14. Gilles Deleuze and Félix Guattari, *What Is Philosophy?* trans. Hugh Tomlinson and Graham Burchell (New York: Columbia University Press, 1994), 42.

15. Ibid., 202.

16. Ibid.

17. Ibid., 201.

18. Ibid.

19. Ibid., 202.

20. Ibid.

21. Ibid., 204.

22. Ibid., 206.

23. Ibid., 204.

24. Zourabichvili, *Deleuze*, 193.

25. Félix Guattari, *The Anti-Oedipus Papers*, trans. Kélina Gotman (New York: Semitext(e), 2006), 401.

26. Ibid.

27. Gilles Deleuze, *Negotiations*, trans. Martin Joughin (New York: Columbia University Press, 1995), 14.

28. Véronique Bergen, "Topologie de la précarité chez Gillez Deleuze et Alain Badiou," lecture given at the conference "Experimenting with Intensities: Science, Philosophy, Politics, the Arts," Peterborough, Ontario, Canada, Trent University, May 12–15, 2004.

29. Deleuze and Guattari, *What Is Philosophy?* 91.

30. Deleuze and Guattari, *A Thousand Plateaus*, 25.

31. In this sense, a non-system or anti-system is still a system.

32. One could give to the word "exhaust" the very sense that Deleuze confers upon it when speaking about Beckett.

33. [Martin Heidegger, *What Is Called Thinking?* trans. J. Glenn Gray (New York: Harper and Row, 1968), 8.—Trans.]

34. Deleuze's famous phrase—quoting Kierkegaard—"possibility, lest I suffocate" was spoken in a lecture given on May 31, 1983, in his course titled "Cinema: A Classification of Signs and Time."

35. Nancy, *The Muses*, 5. Translation amended.

36. Ibid., 5.

37. Ibid., 26. Translation amended.

38. Ibid., 37.

39. Here we are referring to "The Question Concerning Technology" (1954), which may be found in Martin Heidegger, *Basic Writings*, trans. David

Farrell Krell (New York: HarperCollins, 1993), 303–41. A general yet very precise study—particularly concerning the connection between *deinon*, *technē*, and according to Heidegger—may be found in Emilio Brito, *Heidegger et l'hymne du sacré* (Leuven: Leuven University Press, 1999).

40. Heidegger, "The Question Concerning Technology," 313.

41. Ibid., 337.

42. Martin Heidegger, *Introduction to Metaphysics*, trans. Gregory Fried and Richard Polt (New Haven, Conn.: Yale University Press, 2000), 10.

43. Ibid., 13.

44. Ibid.

45. Ibid., 14.

46. Ibid., 5–6.

47. Friedrich Hölderlin, *Selected Poems of Friedrich Hölderlin*, trans. Maxine Chernoff and Paul Hoover (Richmond, Calif.: Omnidawn, 2008), 283.

48. Jacques Desautels, for example (and this example is quite a general one), does not even devote a line to the *Timaeus* in his chapter "Les cosmologies grecques" in his voluminous work, *Dieux et mythes de la Grèce ancienne* (Quebec: Presses de l'Université Laval, 1988).

49. *Timaeus* 30a. [Barrau closely follows the translation and translator notes of Luc Brisson in *Timée*, trans. Luc Brisson (Paris: Flammarion, 1992). At times, however, Brisson's translation of the *Timaeus* departs from standard English translations. When no French is given, all English translations of the *Timaeus* are taken from: Plato, *Plato's Timaeus*, trans. Peter Kalkavage (Newburyport, Mass.: Focus, 2001), 60. When the French is given alongside the English translation above, it indicates an English translation of Brisson's French translation. Finally, Stephanus numbers have been included for readers who would like to compare the translations of Brisson and Kalkavage alongside the original Greek.—Trans.]

50. *Timaeus* 29d. Luc Brisson underscores that because he gave an explanatory sense to *kai*, he had no other choice but to identify becoming with the universe in his translation.

51. In reality, there is no possessive adjective in the Greek text. Luc Brisson's translation and the point that we have underscored are not incorrect insofar as it is indeed "this world" that is designated.

52. *Timaeus* 30b. [Here, and elsewhere, Brisson translates *kosmos* as *monde* or world.—Trans.]

53. *Timaeus* 41a. Translation of Donald Zeyl, "Timaeus," in *Plato: Complete Works*, ed. John M. Cooper (Indianapolis, Ind.: Hackett, 1997), 1244.

54. *Timaeus* 31a.

55. *Timaeus* 31b.

56. See Robert William Jordan, *Plato's Arguments for Forms*, in *Proceedings of the Cambridge Philosophical Society*, supplementary vol. 9 (1983), and the

summary offered by Luc Brisson in his introduction to the fifth edition of the *Timaeus* (Paris: Flammarion, 2001).

57. Each "definition" or term here could be replaced by its opposite. *Khōra* is in fact undefinable.

58. Jean-Pierre Vernant, *Myth and Society in Ancient Greece*, trans. Janet Lloyd (New York: Zone Books, 1980), 240.

59. Derrida, "Khōra."

60. Ibid., 89.

61. Ibid., 96.

62. Ibid., 90.

63. Ibid., 92.

64. Ibid., 93.

65. Ibid., 93–94.

66. Ibid., 94.

67. Ibid., 95.

68. Ibid., 96.

69. Ibid., 101.

70. Ibid., 103.

71. Ibid.

72. Ibid., 104.

73. Ibid., 112.

74. Ibid., 121.

75. Ibid., 119.

76. Ibid., 120.

77. Jacques Derrida, "Structure, Sign, and Play," in *Writing and Difference*, trans. Alan Bass (Chicago: University of Chicago Press, 1978), 288.

78. [Cf. p. 49—Trans.]

79. Hilary Putnam, *Renewing Philosophy* (Cambridge, Mass.: Harvard University Press, 1992), 133.

80. Jean-Luc Nancy, *Dis-Enclosure: The Deconstruction of Christianity*, trans. Bettina Bergo, Gabriel Malenfant, and Michael B. Smith (New York: Fordham University Press, 2008), 15.

81. [Cf. p. 58—Trans.]

82. Jean-Luc Nancy, *The Sense of the World*, trans. Jeffrey S. Librett (Minneapolis: University of Minnesota Press, 1997), 166.

83. Ibid., 77.

84. Ibid., 79.

85. Ibid., 4.

86. [Cf. p. 57—Trans.]

87. Cf., for example, Jacques Derrida, *The Post Card: From Socrates to Freud and Beyond*, trans. Alan Bass (Chicago: University of Chicago Press, 1987).

88. Here the question is clearly not to deny the legitimacy or the validity of cosmology (especially physical ones) or to restrict in any way its developments to come. These developments are as anticipated as they are necessary. Rather the question is simply to underscore that, according to the mode of struction, an entanglement of other worlds—which are no longer entirely worlds—must *also* be thought.

89. Alberto Caeiro and Fernando Pessoa, *The Keeper of Sheep*, trans. Edwin Honig and Susan M. Brown (Riverdale-on-Hudson, N.Y.: Sheep Meadow Press, 1986).

90. Richard Rorty, *Philosophy and the Mirror of Nature* (Princeton, N.J.: Princeton University Press, 1979), 4.

91. Bernard d'Espagnat, *Veiled Reality: An Analysis of Quantum Mechanical Concepts* (Boulder, Colo.: Westview Press, 2003).

92. Hélène Cixous and Jacques Derrida, *Veils*, trans. Geoffrey Bennington (Stanford, Calif.: Stanford University Press, 2001).

93. Jacques Derrida, *Dissemination*, trans. Barbara Johnson (Chicago: University of Chicago Press, 1981), 71.

94. Cixous and Derrida, *Veils*, 3–13.

95. [Here Barrau cites Derrida's untranslatable phrase *coup de don*. The phrase plays on the common idiomatic phrase structure in French, *coup de*. What is interesting, however, is that whereas the French reader expects to hear one of the many conventional or idiomatic phrases with *coup de*, such as a "stroke of luck" (*coup de chance*) or a "final blow" (*coup de grâce*), here Derrida inserts the highly unconventional and atypical word "gift" or *don* after *coup de*. The word, however, does not belong in the idiom. It disturbs, even perturbs, the French ear and the French language. In other words, it interrupts one language with another language that is outside of it.—Trans.]

96. Jacques Derrida, *Spurs: Nietzsche's Styles/Eperons: Les Styles de Nietzsche*, trans. Barbara Harlow (Chicago: University of Chicago Press, 1979), 59.

97. Ibid., 71.

98. [To the French-speaking ear, the phrase *plier voile* collages two typical French expressions—*plier bagage* (to pack your things quietly and leave) and *mettre les voiles* (to raise the sails and leave unnoticed)—both of which mean "to leave discreetly."—Trans.]

99. Jean-Luc Nancy, *Listening*, trans. Charlotte Mandell (New York: Fordham University Press, 2007), 7.

100. Bruno Pinchard mentions this incredible quote of Mozart: "I do not hear in my imagination each part successively, but rather I hear them at the same time all together" (Pinchard, "Polyphonie du désordre," http://facdephilo.univ-lyon3.fr/servlet/com.univ.collaboratif.utils.LectureFichiergw?ID_FICHIER=1287570601157).

Stefanos Geroulanos and Todd Meyers, *series editors*

Georges Canguilhem, *Knowledge of Life.* Translated by Stefanos Geroulanos and Daniela Ginsburg, Introduction by Paola Marrati and Todd Meyers.

Henri Atlan, *Selected Writings: On Self-Organization, Philosophy, Bioethics, and Judaism*. Edited and with an Introduction by Stefanos Geroulanos and Todd Meyers.

Catherine Malabou, *The New Wounded: From Neurosis to Brain Damage.* Translated by Steven Miller.

François Delaporte, *Chagas Disease: History of a Continent's Scourge.* Translated by Arthur Goldhammer, Foreword by Todd Meyers.

Jonathan Strauss, *Human Remains: Medicine, Death, and Desire in Nineteenth-Century Paris.*

Georges Canguilhem, *Writings on Medicine.* Translated and with an Introduction by Stefanos Geroulanos and Todd Meyers.

François Delaporte, *Figures of Medicine: Blood, Face Transplants, Parasites.* Translated by Nils F. Schott, Foreword by Christopher Lawrence.

Juan Manuel Garrido, *On Time, Being, and Hunger: Challenging the Traditional Way of Thinking Life.*

Pamela Reynolds, *War in Worcester: Youth and the Apartheid State.*

Vanessa Lemm and Miguel Vatter, eds., *The Government of Life: Foucault, Biopolitics, and Neoliberalism.*

Henning Schmidgen, *The Helmholtz Curves: Tracing Lost Time.* Translated by Nils F. Schott.

Henning Schmidgen, *Bruno Latour in Pieces: An Intellectual Biography.* Translated by Gloria Custance.

Veena Das, *Affliction: Health, Disease, Poverty.*

Kathleen Frederickson, *The Ploy of Instinct: Victorian Sciences of Nature and Sexuality in Liberal Governance.*

Roma Chatterji, ed., *Wording the World: Veena Das and Scenes of Inheritance.*

Jean-Luc Nancy and Aurélien Barrau, *What's These Worlds Coming To?* Translated by Travis Holloway and Flor Méchain. Foreword by David Pettigrew.